The Scie
Case for Creation

Bert Thompson, Ph.D.

APOLOGETICS PRESS

Apologetics Press, Inc.
230 Landmark Drive
Montgomery, Alabama 36117-2752

First Edition © Copyright 1986
First Revised Edition © Copyright 1999
Second Revised Edition © Copyright 2002
Third Revised Edition © Copyright 2004
ISBN: 0-932859-63-1

TABLE OF CONTENTS

1

INTRODUCTION

There are two fundamentally different, and diametrically opposed, explanations for the origin of the Universe, the origin of life in that Universe, and the origin of new types of varying life forms. Each of these explanations is a cosmogony—an entire world view, or philosophy, of origins and destinies, of life and meaning.

One of these world views is the concept of **evolution**. According to the theory of evolution, or as it may be called more properly, the evolution model, the Universe is **self-contained**. Everything in the Universe has come into being through mechanistic processes without any kind of supernatural intervention. This view asserts that the origin and development of the Universe and all of its systems (the Universe itself, living non-human organisms, man, etc.) can be explained solely on the basis of time, chance, and continuing natural processes innate in the structure of matter and energy.

According to this particular theory, all living things have arisen from a single-celled organism, which in turn had arisen from an inanimate, inorganic world. This theory may be called the "General Theory of Evolution," a name given to it by G.A. Kerkut, the famous British evolutionist/physiologist who described it as "...the theory that all the living forms in the world have arisen from a single source which itself came from an inorganic form" (1960, p. 157).

The second alternate and opposing world view is the concept of **creation**. According to the theory of creation, or as it may be called more properly, the creation model, the Universe is **not self-contained**. Everything in the Universe, and in fact, the Universe itself, has come into being through the design, purpose, and deliberate acts of a supernatural Creator Who, using processes that are not continuing as natural processes in the present, created the Universe, the Earth, and all life on the Earth, including all basic types of plants and animals, as well as humans.

As various authors—both evolutionists (see Wald, 1979, p. 289) and creationists (see Wysong, 1976, p. 5)—have observed, there are two and only two possibilities regarding origins. One or the other of these two philosophies (or models) must be true. That is to say, all things either can, or cannot, be explained in terms of ongoing natural processes in a self-contained Universe. If they can, then evolution is true. If they cannot, then they must be explained, at least in part, by extranatural processes that can account for a Universe which itself was created. In their text, *What Is Creation Science?*, Henry Morris and Gary Parker commented on this point.

> The fact is, however, there are **only two** possible models of origins, evolution or creation.... Either the space/mass/time universe is eternal, or it is not. If it is, then evolution is the true explanation of its various components. If it is not, then it must have been created by a Creator. These are the only two possibilities—simply stated, either it happened by accident (chance)... **or it didn't** (design).... There are only these two possibilities. There may be many evolution submodels... and various creation submodels..., but there can be only two basic models—evolution or creation (1987, p. 190, emp. in orig.).

Various terms have been used to describe the two concepts of origins—creation versus evolution, design versus chance, theism versus naturalism/materialism, etc.—but in the end all of these phrases are merely different ways of expressing the same two basic alternatives.

Some, in an attempt to offer a third alternative, have suggested "theistic evolution" (also known as "directed evolution," "mitigated evolution," or "religious evolution"), which postulates both a Creator and an evolutionary scenario. Evolutionists frequently have been known to ask creationists, "Which creation story do you wish to see taught—Buddhist? Hindu? Christian?, etc.?" The fact remains, of course, that ultimately either there is a Creator or there is not. That question will have to be resolved, whether or not one wishes to retreat to a concept like theistic evolution. An appeal to theistic evolution as a possible "third alternative" in the origins controversy will not answer the basic questions involved. Also, evolutionists need to be reminded that the cosmogonies of the Buddhists, Hindus, Taoists, Confucianists, etc. are all based on evolution. Orthodox Jewish, Muslim, and Christian cosmogonies are all based on creation. Anyone who takes the time and expends the effort to study these issues likely will come to realize the illogical, contradictory nature of theistic evolution and related concepts (see Thompson, 1977, 1995, 2000). There may be many evolutionary submodels (e.g., different mechanisms, rates, or sequences) and various creationist submodels (e.g., different dates, or events of creation), but there still remain only two basic models—creation and evolution.

Both evolution and creation may be referred to correctly as scientific models, since both may be used to explain and predict scientific facts. Obviously the one that does the better job of explaining/predicting is the better scientific model. However, by the very nature of how science works, simply because one model fits the facts better does not prove it true. Rather, the model that better fits the available scientific data is said to be the one that has the highest degree of probability of being true. Knowledgeable scientists understand this, of course, and readily accept it, recognizing the limitations of the scientific method (due to its heavy dependence upon inductive, rather than strictly deductive, reasoning).

In order to examine properly the two models, they must be defined in broad, general terms, and then each must be compared to the available data in order to examine its effectiveness in explaining and predicting various scientific facts. What, then, by way of summary, do the two different models predict and/or include? The **evolution model** includes the evidence from various fields of science for a gradual emergence of present life kinds over eons of time, with emergence of complex and diversified kinds of life from "simpler" kinds, and ultimately from nonliving matter. The **creation model** includes the evidence from various fields of science for a sudden creation of complex and diversified kinds of life, with gaps persisting between different kinds, and with genetic variation occurring within each kind. The creation model denies "vertical" evolution (also called "macroevolution"—the emergence of complex from simple, and change between kinds), but does not challenge "horizontal" evolution (also called "microevolution"—the formation of species or subspecies within created kinds, or genetic variation). In defining the concepts of creation and evolution, an examination of several different aspects of each of the models demonstrates the dichotomy between the two. Put into chart form, such a comparison would appear as seen in Table 1 on the next page.

The creation model includes the scientific evidence and the related inferences suggesting that:	The evolution model includes the scientific evidence and the related inferences suggesting that:
I. The Universe and the solar system were created suddenly.	I. The Universe and the solar system emerged by naturalistic processes.
II. Life was created suddenly.	II. Life emerged from non-life via naturalistic processes.
III. All present living kinds of animals and plants have remained fixed since creation, other than extinctions, and genetic variation in originally created kinds has occurred only within narrow limits.	III. All present kinds emerged from simpler earlier kinds, so that single-celled organisms evolved first into invertebrates, then vertebrates, then amphibians, then reptiles, then mammals, then primates (including man).
IV. Mutation and natural selection are insufficient to have brought about the emergence of present living kinds from a simple primordial organism.	IV. Mutation and natural selection have brought about the emergence of present complex kinds from a simple primordial organism.
V. Man and apes have a separate ancestry.	V. Man and apes emerged from a common ancestor.
VI. The Earth's geologic features appear to have been fashioned largely by rapid, catastrophic processes that affected the Earth on a global and regional scale (catastrophism).	VI. The Earth's geologic lectures were fashioned largely by slow, gradual processes, with infrequent catastrophic events restricted to a local scale (uniformitarianism).
VII. The inception of both the Earth and living kinds may have been relatively recent.	VII. The inception of both the Earth and of life must have occurred several billion years ago.

Table 1 – The two models of origins (after Gish, et al., 1981)

2

IMPORTANCE OF THE CREATION/EVOLUTION CONTROVERSY

The creation/evolution question is hardly a trivial issue that concerns only a few scientists on the one hand or a few religionists on the other. In one way or another, the issue permeates practically every field of academic study and every aspect of national life. It deals with two opposing world views. Consequently, it should be of interest to almost everyone. Certainly, few would doubt that in recent years the controversy definitely has heightened. Various states have discussed enacting, or have attempted to enact, laws that militate against the teaching of the scientific evidence of only one theory of origins. Books are being written by evolutionists that attack the creationist stance; books are being written by creationists that attack the evolutionist stance. National news media have become involved. Science associations have become involved. Teachers' associations and political groups have become involved. Far from diminishing, the controversy seems to be increasing. And both sides acknowledge that it is not likely to "go away." As one evolutionist put it in commenting on the upswing of creationism in America: "The climate of the times suggests that the problem will be with us for a very long time..." (Moore, 1981, p. 1). Indeed, "the problem" will be with us for a very long time.

There was a time when creationists and their arguments largely were ignored by many in the scientific community. That hardly is the case now, however. And there is good reason why evolutionary scientists have become alarmed enough to consider creation a threat.

In 1971, Harvard-trained lawyer Norman Macbeth wrote a biting rebuttal of evolution titled *Darwin Retried*. Somewhat later, in a published interview about the book and its contents, he observed that evolutionists were "not revealing all the dirt under the rug in their approach to the public. There is a feeling that they ought to keep back the worst so that their public reputation would not suffer and the Creationists wouldn't get any ammunition" (1982, 2:22). It is too late, however, because the evolutionists' public reputation **has** suffered, and the creationists **have** garnered to themselves additional ammunition, as is evident from the following.

In a center-column, front-page article in the June 15, 1979 issue of the *Wall Street Journal,* there appeared an article by one of the *Journal's* staff writers commenting on how creationists, when engaging in debates with evolutionists, "tend to win" the debates, and that creationism was "making progress." In 1979, Gallup pollsters conducted a random survey in America, inquiring about belief in creation versus evolution. The poll had been commissioned by *Christianity Today* magazine, and was reported in its December 21, 1979 issue. This poll found that 51% of Americans believe in the special creation of a literal Adam and Eve as the starting place of human life. A 1980 Gallup poll showed that over half of the United States population believed in a literal, specially created Adam and Eve as the parents of the whole human race. The March 1980 issue of the *American School Board Journal* (p. 52) announced that 67% of its readers (most of whom were school board members and school administrators) favored the teaching of the scientific evidence for creation in public schools. One of the most authoritative polls was conducted in October 1981 by the Associated Press/NBC News polling organization. The results were as follows:

"Only evolution should be taught"	8%
"Only creation should be taught	10%
"Both creation & evolution should be taught"	76%
"Not sure which should be taught"	6%

Thus, nationwide no less than 86% of the people in the United States believe that creation should be taught in public schools. In August 1982, another Gallup poll was conducted and found that 44% (i.e., almost half) of the population believed not only in creation, but in a recent creation occurring less than 10,000 years ago (see Morris, 1982b, pp. 12,130,164; also see *San Diego Union*, 1982). *Glamour* magazine conducted a poll of its own, and reported the results in its August 1982 issue (p. 28). The magazine found that 74% of its readers favored teaching the scientific evidence for creation in public schools.

Amazingly, after almost a decade (and in some cases more than a decade), these figures have changed very little. On November 28, 1991, results were released from yet another Gallup poll regarding the biblical account of origins. The results may be summarized as follows. On origins: 47% believed God created man within the last 10,000 years (up 3% from the 1982 poll mentioned above); 40% believed man evolved over millions of years, but that God guided the process; 9% believed man evolved over millions of years without God; 4% were "other/don't know." On the Bible: 32% believed the Bible to be the inspired Word of God and that it should be taken literally; 49% believed the Bible to be the inspired Word of God, but that it should not always be taken literally; 16% believed the Bible to be entirely the product of men; 3% were "other/don't know" (see Major, 1991a, 11:48; John Morris, 1992, p. d). Two years later, a Gallup poll carried out in 1993 produced almost the same results. Of those responding, 47% stated that they believed in a recent creation of man; 11% expressed their belief in a strictly naturalistic form of evolution (see Newport, 1993, p. A-22). Four years after that poll, a 1997 Gallup survey found that 44% of Americans (including 31% who were college graduates) subscribed to a fairly literal reading of the Genesis account of creation, while another 39% (53% of whom were college graduates) believed God played at least some part in creating the Universe. Only 10% (17% college graduates) embraced a purely naturalistic, evolutionary view (see Bishop, 1998, pp. 39-48; Sheler, 1999, pp. 48-49). The results of a Gallup poll released in August 1999 were practically identical: 47% stated that they believed in a recent creation of man; 9% expressed belief in strictly naturalistic evolution (see Moore, 1999).

In its March 11, 2000 issue, the *New York Times* ran a story titled "Survey Finds Support is Strong for Teaching 2 Origin Theories," which reported on a poll commissioned by the liberal civil rights group, People for the American Way, and conducted by the prestigious polling/public research firm, DYG, of Danbury, Connecticut. According to the report, 79% of the people polled felt that the scientific evidence for creation should be included in the curriculum of public schools (see Glanz, 2000, p. A-1).

These results were unexpected by evolutionists, who would have expected instead a general agreement with evolutionary theory in light of the many decades of indoctrination in the schools, textbooks, and news media to the effect that evolution is a "fact" and that the Earth is billions of years old. Little wonder, then, that many evolutionists are becoming alarmed regarding the creationist position.

EVOLUTIONARY SCIENTISTS AS "RELUCTANT CREATIONISTS"?

No doubt the shock that so many today believe in the concept of creation is devastating news to evolutionists. But now, as if to add salt to an already open and bleeding wound, some in the evolutionary camp are "defecting" as well. Gary Parker, in the section of *What Is Creation Science?* that he authored, stated:

> The case for **creation**, however, is not based on imagination. Creation is based instead on **logical inference** from our **scientific observations**, and on simple acknowledgment that everyone, scientists and laymen alike, recognize that certain kinds of design imply creation.... According to creation, living things **operate** in understandable ways that can be described

in terms of scientific laws—but these observations include properties of organization that logically imply a created origin for life.

The creationist, then, recognizes the orderliness that the vitalist doesn't see. But he doesn't limit himself only to those kinds of order that result from time, chance, and the properties of matter as the evolutionist does. Creation introduces levels of order and organization that greatly enrich the range of explorable hypotheses and turn the study of life into a scientist's dream.

If the evidence for the creation of life is as clear as I say it is, then other scientists, even those who are evolutionists, ought to see it—and they do (Morris and Parker, 1987, p. 47, emp. in orig.).

They do? Even evolutionists? Apparently so. Consider, for example, the following. On November 5, 1981, the late Colin Patterson, who was serving at the time as the senior paleontologist at the British Museum of Natural History in London, and who was recognized widely as one of the world's foremost evolutionary experts, delivered an address to his evolutionist colleagues at the American Museum of Natural History in New York. In that speech, Dr. Patterson astonished those assembled by stating that he had been "kicking around" non-evolutionary, or "anti-evolutionary," ideas for approximately eighteen months. As he described it:

One morning I woke up and something had happened in the night, and it struck me that I had been working on this stuff for twenty years and there was not one thing I knew about it. That's quite a shock to learn that one can be misled so long. Either there was something wrong with me, or there was something wrong with evolution theory (1981).

Dr. Patterson said he knew there was nothing wrong with him, so he started asking various individuals and groups a simple question: "Can you tell me anything you know about evolution, any one thing that is true? I tried that question on the geology staff at the Field Museum of Natural History, and the only answer I got was silence." He then tried the same tactic with people in attendance at an evolutionary morphology seminar at the University of Chicago (a very prestigious body of evolutionists), and all he got there, according to his personal report of the event, "was silence for a long time and eventually one person said, 'I know one thing—it ought not to be taught in high school.'" He then remarked, "It does seem that the level of knowledge about evolution is remarkably shallow. We know it ought not to be taught in high school, and that's all we know about it."

Patterson went on to say: "Then I woke up and realized that all my life I had been duped into taking evolution as revealed truth in some way." But even more important, he termed evolution an "anti-theory" that produced "anti-knowledge." He also suggested that "the explanatory value of the hypothesis is nil" and that evolution theory is "a void that has the function of knowledge but conveys none." To use Patterson's wording, "I feel that the effect of hypotheses of common ancestry in systematics has not been merely boring, not just a lack of knowledge, I think it has been positively anti-knowledge" (1981).

Dr. Patterson made it clear, as I wish to do here, that he never had any fondness for the creationist position. Yet he was willing to label his stance as "anti-evolutionary," which was quite a change for a man who had authored several books in the field he eventually came to believe produces nothing but "anti-knowledge."

Colin Patterson was not the only scientist who expressed such views. For more than two decades, the late, distinguished British astronomer Sir Fred Hoyle stressed the serious problems, especially from the fields of thermodynamics, with theories about the naturalistic origin of life on the Universe. In 1981, Dr. Hoyle wrote:

> I don't know how long it is going to be before astronomers generally recognize that the combinatorial arrangement of not even one among the many thousands of biopolymers on which life depends could have been arrived at by natural processes here on the Earth. Astronomers will have a little difficulty in understanding this because they will be assured by biologists that it is not so, the biologists having been assured in their turn by others that it is not so. The "others" are a group of

persons who believe, quite openly, in mathematical miracles. They advocate the belief that tucked away in nature, outside of normal physics, there is a law which performs miracles (provided the miracles are in the aid of biology). This curious situation sits oddly on a profession that for long has been dedicated to coming up with logical explanations of biblical miracles.... It is quite otherwise, however, with the modern miracle workers, who are always to be found living in the twilight fringes of thermodynamics (1981a, p. 526).

In fact, Dr. Hoyle went on to remark:

The likelihood of the spontaneous formation of life from inanimate matter is one to a number with 40,000 noughts after it.... It is big enough to bury Darwin and the whole theory of evolution. There was no primeval soup, neither on this planet nor on any other, and if the beginnings of life were not random, they must therefore have been the product of purposeful intelligence (1981b, 294:148).

He then described the evolutionary concept that disorder gives rise to order in a rather picturesque manner. He said that "the chance that higher forms have emerged in this way is comparable with the chance that a tornado sweeping through a junkyard might assemble a Boeing 747 from the materials therein" (1981b, 294:105). To make his position perfectly clear, he provided his readers with the following analogy:

At all events, anyone with even a nodding acquaintance with the Rubik cube will concede the near-impossibility of a solution being obtained by a blind person moving the cubic faces at random. Now imagine 10^{50} blind persons each with a scrambled Rubik cube, and try to conceive of the chance of them all **simultaneously** arriving at the solved form. You then have the chance of arriving by random shuffling at just one of the many biopolymers on which life depends. The notion that not only biopolymers but the operating programme of a living cell could be arrived at by chance in a primordial organic soup here on the Earth is evidently nonsense of a high order (1981a, p. 527, emp. in orig.).

Hoyle, and his colleague Chandra Wickramasinghe (professor of astronomy and applied mathematics at University College, Cardiff, Wales), employed probabilistic statistics (applied to cosmic time, not just geologic time here on Earth) to investigate the possibility of the naturalistic origin of life, and concluded:

> Once we see, however, that the probability of life originating at random is so utterly minuscule as to make the random concept absurd, it becomes sensible to think that the favourable properties of physics on which life depends, are in every respect deliberate.... It is therefore almost inevitable that our own measure of intelligence must reflect in a valid way the higher intelligences...even to the extreme idealized limit of **God** (1981, pp. 141,144, emp. in orig.).

Hoyle and Wickramasinghe suggested, however, that this "higher intelligence" does not necessarily have to be, as far as they are concerned, what most people would call "God," but a being with an intelligence "even to the limit of God." They opted instead for a "directed panspermia," which suggests that life was "planted" on Earth, through genetic material, by a "higher intelligence" somewhere in the Universe.

The point I wish to make here is that even scientists who are not creationists are able to recognize that creation is a **legitimate scientific concept** whose merits deserve to be compared with those of evolution. And some make statements that at least lean more toward the scientific respectability of creation than toward that of evolution. For example, a thought-provoking article by British physicist H.S. Lipson appeared in the May 1980 issue of *Physics Bulletin*. In his article, "A Physicist Looks at Evolution," Dr. Lipson commented first on his interest in life's origin, and second on his non-association with any type of creation theory, but then noted: "In fact, evolution became in a sense a scientific religion; almost all scientists have accepted it, and many are prepared to 'bend' their observations to fit with it." Dr. Lipson then asked how well evolution has withstood years of scientific testing, and suggested that "to my mind, the theory does not stand up at all."

After reviewing many of the problems (especially from thermodynamics) that would be involved in producing something living from something nonliving, he asked: "If living matter is not, then, caused by the interplay of atoms, natural forces, and radiation, how has it come into being?" Dr. Lipson dismissed any sort of "directed evolution" (a British term for what people in America generally refer to as "theistic evolution"), and concluded: "I think, however, that we must go further than this and admit that the only acceptable explanation is **creation**." Like Hoyle, Wickramasinghe, and Patterson, Dr. Lipson is not happy about the conclusion he has been forced to draw from the evidence. He made that clear when he said: "I know that this is anathema to physicists, as indeed it is to me, but we must not reject a theory that we do not like if the experimental evidence supports it" (1980, 31:138, emp. in orig.).

Interestingly, just two years before Dr. Lipson penned his article, Harvard geneticist Richard Lewontin made the following comment in the September 1978 issue of *Scientific American*, which was devoted in its entirety to a defense of organic evolution:

> Life forms are more than simply multiple and diverse, however. Organisms fit remarkably well into the external world in which they live. They have morphologies, physiologies and behaviors that **appear to have been carefully and artfully designed** to enable each organism to appropriate the world around it for its own life. It was the marvelous fit of organisms to the environment, much more than the great diversity of forms, that was the chief evidence of a **Supreme Designer** (1978, 239[3]:213, emp. added).

Of course, Dr. Lewontin then went on to try to explain in his article how nature alone—without any assistance whatsoever from a "Supreme Designer"—could account for the impressive "apparent design" in the world around us.

Three years before Dr. Lipson wrote his article, France's preeminent zoologist, Pierre-Paul Grassé (whose knowledge of the living world has been called by his colleagues "encyclopedic"), authored *The Evolution of Living Organisms,* in which he wrote:

Today our duty is to destroy the myth of evolution, considered as a simple, understood, and explained phenomenon which keeps rapidly unfolding before us. Biologists must be encouraged to think about the weaknesses and extrapolations that theoreticians put forward or lay down as established truths. The deceit is sometimes unconscious, but not always, since some people, owing to their sectarianism, purposely overlook reality and refuse to acknowledge the inadequacies and falsity of their beliefs.

Their success among certain biologists, philosophers, and sociologists notwithstanding, **the explanatory doctrines of biological evolution do not stand up to an objective, in-depth criticism**. They prove to be either in conflict with reality, or else incapable of solving the major problems involved (1977, pp. 8,202, emp. added).

Five years after Lipson's statements, Michael Denton authored his classic text, *Evolution: A Theory in Crisis*, and remarked:

In this book, I have adopted the radical approach. By presenting a systematic critique of the current Darwinian model, ranging from paleontology to molecular biology, I have tried to show why I believe that the problems are too severe and too intractable to offer any hope of resolution in terms of the orthodox framework, and that consequently the conservative view is no longer tenable.

The intuitive feeling that pure chance could never have achieved the degree of complexity and ingenuity so ubiquitous in nature has been a continuing source of scepticism ever since the publication of *Origin*; and throughout the past century there has always existed a significant minority of first-rate biologists who have never been able to bring themselves to accept the validity of Darwin's claims. In fact, the number of biologists who have expressed some degree of disillusionment is practically endless.

The anti-evolution thesis argued in this book, the idea that life might be fundamentally a discontinuous phenomenon, runs counter to the whole thrust of biological thought.... Put simply, no one has ever observed the interconnecting continuum of functional forms linking all known past and present species of life. The concept of continuity of nature has existed in the mind of man, **never** in the facts of nature (1985, pp. 16,327, 353, emp. in orig).

A year later, when Oxford University's renowned evolutionist Richard Dawkins published *The Blind Watchmaker*, he lamented in the preface: "The complexity of living organisms is matched by the elegant efficiency of their apparent design. **If anyone doesn't agree that this amount of complex design cries out for an explanation, I give up!**" (1986, emp. added). One year after that, the highly regarded Swedish biologist, Søren Løvtrup, wrote:

> After this step-wise elimination, only one possibility remains: **the Darwinian theory of natural selection**, whether or not coupled with Mendelism, **is false**. I have already shown that the arguments advanced by the early champions were not very compelling, and that there are now considerable numbers of empirical facts which do not fit with the theory. Hence, **to all intents and purposes the theory has been falsified**, so why has it not been abandoned? I think the answer to this question is that current evolutionists follow Darwin's example—they refuse to accept falsifying evidence (1987, p. 352, emp. added).

Again, one year later, American physicist George Greenstein wrote in his book, *The Symbiotic Universe:*

> As we survey all the evidence, the thought insistently arises that some supernatural agency—or, rather, Agency—must be involved. Is it possible that suddenly, without intending to, we have stumbled upon scientific proof of the existence of a Supreme Being? Was it God who stepped in and so providentially crafted the cosmos for our benefit? (1988, p. 27).

[Greenstein quickly went on to voice his dissent with such a conclusion, which he considered a "heady prospect" that he labeled as "illusory" (pp. 27,38).]

These quotations—and in chapter 3 I will provide several more recent examples almost identical to them—are not from creationists. Rather, they are from highly respected evolutionists who are well known for their vigilant support of evolutionary theory. Yet even though the authors of these statements are evolutionists, something has caused them to see that evolution simply is not an adequate explanation, and that the Universe and the life it contains "appear to have been designed"—which is my reason for quoting them here. I do not mention them to suggest that they are creationists. I mention them to document the fact that there are highly respected, well-known non-creationist scientists who are beginning to recognize inescapable evidence of actual (not just "apparent") design in nature. These same scientists have expressed serious doubts in regard to evolutionary concepts that were supposed to be able to explain such design, yet obviously have failed to do so. Thus, these scientists now are willing to call into question those concepts —on a strictly scientific basis—and ask questions like, "Have we stumbled upon scientific proof of the existence of a Supreme Being?" Upon observing such an about face, honest inquirers cannot help but acknowledge the point these evolutionary scientists are making (even if unwittingly): one does not get a painting without a painter, a law without a lawgiver, a poem without a poet—**or design without a designer**!

In fact, after over 120 years of Darwinism, rapidly growing numbers of scientists have become convinced that the natural laws and processes that we now know are at work in the Universe absolutely exclude the possibility that the Cosmos could have created itself, and likewise have become convinced that the scientific evidence demonstrates that living things could not, and in fact, did not, arise from lower forms. Such scientists have become convinced that the concept of creation is a much more credible explanation of the evidence related to origins. I invite your attention as we examine a portion of that evidence in the pages that follow.

3

PLAUSIBILITY OF THE CREATION MODEL

Since in origin science (under discussion here) theories do not lend themselves to the principle of falsification as do theories in operation science, they must be investigated and evaluated on the basis of their plausibility. But how, then, does one go about determining whether an origin-science scenario is plausible? Very simply, the principles of **causality** and **uniformity** can be employed. By **cause** we mean the necessary and sufficient condition that alone can explain the occurrence of a given event. By principle of **uniformity** we mean that the kinds of causes that we observe producing certain effects today can be counted on to have produced similar effects in the past. In other words, what we see as an adequate cause in the present, we assume to have been an adequate cause in the past; what we see as an inadequate cause in the present, we assume to have been an inadequate cause in the past. Evolutionists often have relied on the principles of causality and uniformity in attempts to work out evolutionary scenarios. Thaxton, Bradley, and Olsen have addressed these points.

> Consider, for example, the matter of accounting for the informational molecule, DNA. We have observational evidence in the **present** that intelligent investigators can (and do) build contrivances to channel energy down nonrandom chemical pathways to bring

about some complex chemical synthesis, even gene building. May not the principle of uniformity then be used in a broader frame of consideration to suggest that DNA had an intelligent cause at the beginning? Usually the answer given is no. But theoretically, at least, it would seem the answer should be yes in order to avoid the charge that the deck is stacked in favor of naturalism.

We know that in numerous cases, certain effects always have intelligent causes, such as dictionaries, sculptures, machines and paintings. We reason by analogy that similar effects have intelligent causes. For example, after looking up to see "BUY FORD" spelled out in smoke across the sky, we infer the presence of a skywriter even if we heard or saw no airplane. We would similarly conclude the presence of intelligent activity were we to come upon an elephant-shaped topiary in a cedar forest.

In like manner an intelligible communication via radio signal from some distant galaxy would be widely hailed as evidence of an intelligent source. Why then doesn't the message sequence on the DNA molecule also constitute *prima facie* evidence for an intelligent source? After all, DNA information is not just analogous to a message sequence such as Morse code, it **is** such a message sequence....

We believe that if this question is considered, it will be seen that most often it is answered in the negative simply because it is thought to be inappropriate to bring a Creator into science (1984, pp. 211-212, emp. in orig.).

Use of the principles of uniformity and causality enhance the creation model, for these are cherished concepts of scientific thinking. Albert Einstein once said that scientists are "possessed by the sense of universal causation." Causality confirms that every material effect has an adequate antecedent cause. The basic question, then, is this: Can the origin of the Universe, the origin of life, and the origin of new life forms best be accounted for on the basis of nonintelligent, random, chance, accidental processes? Are these **adequate** causes? Or, are these phenomena best accounted for on the basis of a Creator (i.e., an adequate cause) capable of producing the complex, ordered, information-relating processes we see around us?

What are the options? The Universe exists; therefore, it must be explained in some fashion. However, there are only three ways to account for it: (1) It is **eternal**; (2) It is not eternal; rather it **created itself** from nothing; or (3) It is not eternal, and it did not create itself from nothing; instead, it **was created** by something (or Someone) anterior, and superior, to itself. These three possibilities merit serious attention.

IS THE UNIVERSE ETERNAL?

The front cover of the June 25, 2001 issue of *Time* magazine announced: "How the Universe Will End: Peering Deep Into Space and Time, Scientists Have Just Solved the Biggest Mystery in the Cosmos." Comforting thought, isn't it, to know that the "biggest mystery in the Cosmos" has been figured out? But what, exactly, is that mystery? And why does it merit the front cover of a major news magazine?

The origin and destiny of the Universe always have been important topics in the creation/evolution controversy. In the past, evolutionists went to great extremes to present scenarios that included an eternal Universe, and they went to the same extremes to avoid any scenario that suggested a Universe with a beginning or end because such a scenario posed bothersome questions. In his book, *God and the Astronomers*, the eminent evolutionary astronomer Robert Jastrow, who currently is serving as the director of the Mount Wilson Observatory, put it like this:

> The Universe is the totality of all matter, animate and inanimate, throughout space and time. If there was a beginning, what came before? If there is an end, what will come after? On both scientific and philosophical grounds, the concept of an eternal Universe seems more acceptable than the concept of a transient Universe that springs into being suddenly, and then fades slowly into darkness.

Astronomers try not to be influenced by philosophical considerations. However, the idea of a Universe that has both a beginning and an end is distasteful to the scientific mind. In a desperate effort to avoid it, some astronomers have searched for another interpretation of the measurements that indicate the retreating motion of the galaxies, an interpretation that would not require the Universe to expand. If the evidence for the expanding Universe could be explained away, the need for a moment of creation would be eliminated, and the concept of time without end would return to science. But these attempts have not succeeded, and most astronomers have come to the conclusion that they live in an exploding world (1977, p. 31).

What does Jastrow mean when he says that "these attempts have not succeeded"? And why do evolutionists prefer to avoid the question of a Universe with a beginning? In an interview he granted on June 7, 1994, Dr. Jastrow elaborated on this point. The interviewer, Fred Heeren, asked if there was anything from physics that could explain how the universe first came to be. Jastrow lamented:

No, there's not—this is the most interesting result in all of science.... As Einstein said, scientists live by their faith in causation, and the chain of cause and effect. Every effect has a cause that can be discovered by rational arguments. And this has been a very successful program, if you will, for unraveling the history of the universe. But it just fails at the beginning.... So time, really, going backward, comes to a halt at that point. Beyond that, that curtain can never be lifted.... And that is really a blow at the very fundamental premise that motivates all scientists (as quoted in Heeren, 1995, p. 303).

Seventeen years earlier, in his book, *Until the Sun Dies,* Jastrow had discussed this very problem—a Universe without any adequate explanation for its own existence and, worse still, without any adequate cause for whatever theory scientists might set forth in an attempt to elucidate how it did originate. As Dr. Jastrow noted:

This great saga of cosmic evolution, to whose truth the majority of scientists subscribe, is the product of an act of creation that took place twenty billion years ago [according to evolutionary estimates–BT]. Science, unlike the Bible, has no explanation for the occurrence of that extraordinary event. The Universe, and everything that has happened in it since the beginning of time, are a grand effect without a known cause. An effect without a cause? That is not the world of science; it is world of witchcraft, of wild events and the whims of demons, a medieval world that science has tried to banish. As scientists, what are we to make of this picture? I do not know (1977, p. 21, emp. added).

While Dr. Jastrow may not know **how** the Universe began, there are two things that he and his colleagues **do** know: (1) the Universe had a definite beginning; and (2) the Universe will have a definite ending.

Admittedly, the most comfortable position for the evolutionist is the idea that the Universe is eternal, because it avoids the problem of a beginning or ending and thus the need for any "first cause" such as a Creator. In his book, *Until the Sun Dies*, astronomer Jastrow noted: "The proposal for the creation of matter out of nothing possesses a strong appeal to the scientist, since it permits him to contemplate a Universe without beginning and without end" (1977, p. 32). Jastrow went on to remark that evolutionary scientists preferred an eternal Universe "because the notion of a world with a beginning and an end made them feel so uncomfortable" (p. 33). In *God and the Astronomers*, Dr. Jastrow explained why attempts to prove an eternal Universe had failed miserably. "Now three lines of evidence—the motions of the galaxies, the laws of thermodynamics, and the life story of the stars—pointed to one conclusion; all indicated that the Universe had a beginning" (1978, p. 111). Jastrow—who is considered by many to be one of the greatest science writers of our age—certainly is no creationist. But as a scientist who is an astrophysicist, he has written often on the inescapable conclusion that the Universe had a beginning. Consider, for example, these statements from his pen:

Now both theory and observation pointed to an expanding Universe and a beginning in time.... About thirty years ago science solved the mystery of the birth and death of stars, and acquired new evidence that the Universe had a beginning (1978, pp. 47,105).

[Sir] Arthur Eddington, the most distinguished British astronomer of his day, wrote, "If our views are right, somewhere between the beginning of time and the present day we must place the winding up of the universe." When that occurred, and Who or what wound up the Universe, were questions that bemused theologians, physicists and astronomers, particularly in the 1920's and 1930's (1978, pp. 48-49).

Most remarkable of all is the fact that in science, as in the Bible, the World begins with an act of creation. That view has not always been held by scientists. Only as a result of the most recent discoveries can we say with a fair degree of confidence that the world has not existed forever; that it began abruptly, without apparent cause, in a blinding event that defies scientific explanation (1977, p. 19).

The conclusion to be drawn from the scientific data was inescapable, as Dr. Jastrow himself admitted when he wrote:

The lingering decline predicted by astronomers for the end of the world differs from the explosive conditions they have calculated for its birth, but the impact is the same: **modern science denies an eternal existence to the Universe, either in the past or in the future** (1977, p. 30, emp. added).

In her book, *The Fire in the Equations,* award-winning science writer Kitty Ferguson wrote in agreement.

Our late twentieth-century picture of the universe is dramatically different from the picture our forebears had at the beginning of the century. Today it's common knowledge that all the individual stars we see with the naked eye are only the stars of our home galaxy, the Milky Way, and that the Milky Way is only one among many billions of galaxies. **It's also common knowledge that the universe isn't eternal but had a beginning ten to twenty billion years ago, and that it is expanding** (1994, p. 89, emp. added).

The evidence clearly indicates that the Universe had a beginning. The Second Law of Thermodynamics, as Dr. Jastrow has indicated, shows this to be true. Henry Morris correctly commented: "The Second Law requires the universe to have had a beginning" (1974, p. 26). Indeed, it does. The Universe is not eternal.

Steady State and Oscillating Universe Theories

One theory that was offered in an attempt to establish the eternality of the Universe was the Steady State model, propagated by Sir Fred Hoyle and his colleagues. Even before they offered this unusual theory, however, scientific evidence had been discovered which indicated that the Universe was expanding. Hoyle set forth the Steady State model to: (a) erase any possibility of a beginning; (b) bolster the idea of an eternal Universe; and (c) explain why the Universe was expanding. His idea was that at certain points in the Universe (which he called "irtrons"), matter was being created spontaneously **from nothing**. Since this new matter obviously had to "go" somewhere, and since it is a well-established fact of science that two objects cannot occupy the same space at the same time, it pushed the already-existing matter farther into distant space. Dr. Hoyle asserted that this process of matter continually being created (the idea even came to be known as the "continuous creation" theory) avoided a beginning or ending, and simultaneously accounted for the expansion of the Universe.

For a time, Hoyle's Steady State hypothesis was quite popular. Eventually, however, it was discarded for a number of reasons. Cosmologist John Barrow suggested that the Steady State theory proposed by Hoyle and his colleagues sprang "from a belief that the universe did not have a beginning.... The specific theory they proposed fell into conflict with observation long ago..." (1991, p. 46). Indeed, the Steady State theory did fall into "conflict with observation" for a number of reasons. First, valid empirical observations no longer fit the model (see Gribbin, 1986). Second, new theoretical concepts being proposed were at odds with the Steady State model. Third (and probably most important), the theory violated the First Law of Thermodynamics, which states that neither matter nor energy can be **created** or destroyed in nature. Jastrow commented on this last point when he wrote:

> But the creation of matter out of nothing would violate a cherished concept in science—the principle of the conservation of matter and energy—which states that matter and energy can be neither created nor destroyed. Matter can be converted into energy, and vice versa, but the total amount of all matter and energy in the Universe must remain unchanged forever. It is difficult to accept a theory that violates such a firmly established scientific fact. Yet the proposal for the creation of matter out of nothing possesses a strong appeal to the scientist, since it permits him to contemplate a Universe without beginning and without end (1977, p. 32).

The Steady State model, with its creation of matter from nothing, could not be reconciled with this basic law of science, and thus was abandoned.

Slowly but surely, the Big Bang model of the origin of the Universe eclipsed and eventually replaced the Steady State theory. It postulated that all the matter/energy in the observable Universe was condensed into a particle much smaller than a single proton (the famous "cosmic egg" or "ylem" as it frequently is called). The Big Bang model, however, suffered from at least two major problems. First, it required that whatever made up the "cosmic egg" be eternal—a concept clearly at odds with the Second Law of Thermodynamics. John Gribbin, a highly regarded evolutionary cosmologist, voiced the opinion of many when he wrote: "The biggest problem with the Big Bang theory of the origin of the Universe is philosophical—perhaps even theological—**what was there before the bang**?" (1976, pp. 15-16, emp. added).

Second, the expansion of the Universe could not go on forever; it had to end somewhere. These problems suggested to evolutionists that they were living in a Universe that had a beginning, and that also would have an ending. Robert Jastrow addressed both of these points when he wrote:

And concurrently there was a great deal of discussion about the fact that the second law of thermodynamics, applied to the Cosmos, indicates the Universe is running down like a clock. If it is running down, there must have been a time when it was fully wound up (1978, pp. 48-49).

It was apparent that matter could not be eternal, because, as everyone knows (and as every knowledgeable scientist readily admits), eternal things do not run down. Furthermore, there was going to be an end at some point in the future. And eternal entities do not have either beginnings or endings.

In a desperate effort to avoid any vestige of a beginning or any hint of an ending, evolutionists invented the Oscillating Universe model (also known as the Big Bang/Big Crunch model, the Expansion/Collapse model, etc.). Gribbin suggested that "...the best way round this initial difficulty is provided by a model in which the Universe expands from a singularity, collapses back again, and repeats the cycle indefinitely" (1976, pp. 15-16).

That is to say, there was a Big Bang; but there also will be a Big Crunch, at which time the matter of the Universe will collapse back onto itself. There will be a "bounce," followed by another Big Bang, which will be followed by another Big Crunch, and this process will be repeated ad infinitum. In the Big Bang model, there is a permanent end; not so in the Oscillating Universe model, as Dr. Jastrow explained:

> But many astronomers reject this picture of a dying Universe. They believe that the expansion of the Universe will not continue forever because gravity, pulling back on the outward-moving galaxies, must slow their retreat. If the pull of gravity is sufficiently strong, it may bring the expansion to a halt at some point in the future.
>
> What will happen then? The answer is the crux of this theory. The elements of the Universe, held in a balance between the outward momentum of the primordial explosion and the inward force of gravity, stand momentarily at rest; but after the briefest instant, always drawn together by gravity, they commence to

move toward one another. Slowly at first, and then with increasing momentum, the Universe collapses under the relentless pull of gravity. Soon the galaxies of the Cosmos rush toward one another with an inward movement as violent as the outward movement of their expansion when the Universe exploded earlier. After a sufficient time, they come into contact; their gases mix; their atoms are heated by compression; and the Universe returns to the heat and chaos from which it emerged many billions of years ago (1978, p. 118).

The description provided by Jastrow is that commonly referred to in the scientific literature as the "Big Crunch." But the obvious question after hearing such a scenario is this: After that, then what? Once again, hear Dr. Jastrow:

No one knows. Some astronomers say the Universe will never come out of this collapsed state. Others speculate that the Universe will rebound from the collapse in a new explosion, and experience a new moment of Creation. According to this view, our Universe will be melted down and remade in the caldron of the second Creation. It will become an entirely new world, in which no trace of the existing Universe remains....

This theory envisages a Cosmos that oscillates forever, passing through an infinite number of moments of creation in a never-ending cycle of birth, death and rebirth. It unites the scientific evidence for an explosive moment of creation with the concept of an eternal Universe. It also has the advantage of being able to answer the question: What preceded the explosion? (1978, pp. 119-120).

This, then, is the essence of the Oscillating Universe theory. Several questions arise, however. First, of what benefit would such events be? Second, is such a concept scientifically testable? Third, does current scientific evidence support such an idea?

Of what benefit would a Big Bang/Big Crunch/Big Bang scenario be? **Theoretically**, as I already have noted, the benefit to evolutionists is that they do not have to explain a Universe with an absolute beginning or an absolute ending. A cyclical Universe that infinitely expands and contracts is obviously much more acceptable than one that demands explanations for both its origin and destiny. **Practically**, there is no benefit that derives from such a scenario. The late astronomer from Cornell University, Carl Sagan, noted: "...[I]nformation from our universe would not trickle into that next one and, from our vantage point, such an oscillating cosmology is as definitive and depressing an end as the expansion that never stops" (1979, pp. 13-14).

But is the Oscillating Universe model testable scientifically? Gribbin suggests that it is.

> The key factors which determine the ultimate fate of the Universe are the amount of matter it contains and the rate at which it is expanding.... In simple terms, the Universe can only expand forever if it is exploding faster than the "escape velocity" from itself.... If the density of matter across the visible Universe we see today is sufficient to halt the expansion we can observe today, then the Universe has always been exploding at less than its own escape velocity, and must eventually be slowed down so much that the expansion is first halted and then converted into collapse. On the other hand, if the expansion we observe today is proceeding fast enough to escape from the gravitational clutches of the matter we observe today, then the Universe is and always was "open" and will expand forever (1981, p. 313).

Does the scientific evidence support the theory of an "oscillating," eternal Universe? In the end, the success or failure of this theory depends on two things: (1) the amount of matter contained in the Universe, since there must be enough matter for gravity to "pull back" to cause the Big Crunch; and (2) the amount of gravity available to do the "pulling." The amount of matter required by the theory is one reason why Gribbin admitted: "This, in a nutshell, is one of the biggest problems in cosmology today, the puzzle of the so-called missing mass" (1981, pp. 315-316). Cosmologists, astrophysicists, and astronomers generally refer to the missing mass as "dark matter." In their book, *Wrinkles in Time*, George Smoot and Keay Davidson remarked:

We are therefore forced to contemplate the fact that as much as 90 percent of the matter in the universe is both invisible and quite unknown—perhaps unknowable—to us.... Are such putative forms of matter the fantasies of desperate men and women, frantically seeking solutions to baffling problems? Or are they a legitimate sign that with the discovery of dark matter cosmology finds itself in a terra incognita beyond our immediate comprehension? (1993, pp. 164,171).

In his June 25, 2001 *Time* article (which claims to "solve the biggest mystery in the cosmos"), Michael D. Lemonick dealt with this "puzzle."

As the universe expands, the combined gravity from all the matter within it tends to slow that expansion, much as the earth's gravity tries to pull a rising rocket back to the ground. If the pull is strong enough, the expansion will stop and reverse itself; if not, the cosmos will go on getting bigger, literally forever. Which is it? One way to find out is to weigh the cosmos—to add up all the stars and all the galaxies, calculate their gravity and compare that with the expansion rate of the universe. If the cosmos is moving at escape velocity, no Big Crunch.

Trouble is, nobody could figure out how much matter there actually was. The stars and galaxies were easy; you could see them. But it was noted as early as the 1930s that something lurked out there besides the glowing stars and gases that astronomers could see. Galaxies in clusters were orbiting one another too fast; they should, by rights, be flying off into space like untethered children flung from a fast-twirling merry-go-round. Individual galaxies were spinning about their centers too quickly too; they should long since have flown apart. The only possibility: some form of invisible dark matter was holding things together, and while you could infer the mass of dark matter in and around galaxies, nobody knew if it also filled the dark voids of space, where its effects would not be detectable (2001, 157[25]: 51).

In discussing the Oscillating Universe model, astronomers speak (as Gribbin did in one of the quotes above) of a "closed" or an "open" Universe. If the Universe is **closed**, the Universe will cease its expansion, the Big Crunch could occur (theoretically), and an oscillating Universe becomes (again, theoretically) a viable possibility. If the Universe is **open**, the expansion of the Universe will continue (a condition known as the Big Chill) and the Big Crunch will not occur, making an oscillating Universe impossible. Joseph Silk commented: "The balance of evidence does point to an **open** model of the universe..." (1980, p. 309, emp. added). Gribbin said: "The consensus among astronomers today is that the universe is **open**" (1981, p. 316, emp. added). Jastrow observed: "Thus, the facts indicate that **the universe will expand forever...**" (1978, p. 123, emp. added).

Even more recent evidence seems to indicate that an oscillating Universe is a physical impossibility (see Chaisson, 1992). Evolutionary cosmologist John Wheeler drew the following conclusion based on the scientific evidence available at the time: "With gravitational collapse we come to the end of time. Never out of the equations of general relativity has one been able to find the slightest argument for a 're-expansion' of a 'cyclic universe' or anything other than an end" (1977, p. 15). Astronomer Hugh Ross admitted: "Attempts...to use oscillation to avoid a theistic beginning for the universe all fail" (1991, p. 105). In an article written for the January 19, 1998 issue of *U.S. News and World Report* titled "A Few Starry and Universal Truths," Charles Petit stated:

> For years, cosmologists have wondered if the universe is "closed" and will collapse to a big crunch, or "open," with expansion forever in the cards. **It now seems open—in spades.** The evidence, while not ironclad, is plentiful. Neta Bahcall of Princeton University and her colleagues have found that the distribution of clusters of galaxies at the perceivable edge of the universe imply [sic] that the universe back then was lighter than often had been believed. There appears to be 20 percent as much mass as would be needed to stop the expansion and lead the universe to someday collapse again (124[2]:58, emp. added).

Apparently, the information appearing in the June 25, 2001 *Time* article is "ironclad," and has dealt the ultimate deathblow to the idea of either an eternal or oscillating Universe. In speaking about the origin of the Universe, Lemonick explained:

> That event—the literal birth of time and space some 15 billion years ago—has been understood, at least in its broadest outlines, since the 1960s. But in more than a third of a century, the best minds in astronomy have failed to solve the mystery of what happens at the other end of time. Will the galaxies continue to fly apart forever, their glow fading until the cosmos is cold and dark? Or will the expansion slow to a halt, reverse direction, and send 10 octillion (10 trillion billion) stars crashing back together in a final, apocalyptic Big Crunch, the mirror image of the universe's explosive birth? Despite decades of observations with the most powerful telescopes at their disposal, astronomers simply haven't been able to decide.

> But a series of remarkable discoveries announced in quick succession starting this spring has gone a long way toward settling the question once and for all. Scientists who were betting on a Big Crunch liked to quote the poet Robert Frost: "Some say the world will end in fire,/some say in ice./From what I've tasted of desire/I hold with those who favor fire." Those in the other camp preferred T.S. Eliot: "This is the way the world ends./Not with a bang but a whimper." Now, using observations from the Sloan Digital Sky Survey in New Mexico, the orbiting Hubble Space Telescope, the mammoth Keck Telescope in Hawaii, and sensitive radio detectors in Antarctica, the verdict is in: T.S. Eliot wins (157[25]:49-50).

What, exactly, has caused this current furor in astronomy? And why are T.S. Eliot and the astronomers who quote him the "winners"? As Lemonick went on to explain:

> If these observations continue to hold up, astrophysicists can be pretty sure they have assembled the full parts list for the cosmos at last: 5% ordinary matter, 35% exotic dark matter and about 60% dark energy. They also have a pretty good idea of the universe's future. All the matter put together doesn't have enough gravity to stop the expansion; beyond that, the anti-

gravity effect of dark energy is actually speeding up the expansion. And because the amount of dark energy will grow as space gets bigger, its effect will only increase (157[25]:55).

The simple fact is, the Universe just does not have enough matter, or enough gravity, for it to collapse back upon itself in a "Big Crunch." It is not "oscillating." It is not eternal. It had a beginning, and it will have an ending. As Jastrow observed:

> About thirty years ago science solved the mystery of the birth and death of stars, and acquired new evidence that the Universe had a beginning.... Now both theory and observation pointed to an expanding Universe and a beginning in time" (1978, p. 105).

Six pages later in *God and the Astronomers,* Jastrow concluded: "Now three lines of evidence—the motions of the galaxies, the laws of thermodynamics, the life story of the stars—pointed to one conclusion; all indicated that the Universe had a beginning" (p. 111).

In 1929, Sir James Jeans, writing in his classic book *The Universe Around Us,* observed: "All this makes it clear that the present matter of the universe cannot have existed forever.... In some way matter which had not previously existed, came, or was brought, into being" (1929, p. 316). Now, over seventy years later we have returned to the same conclusion. As Lemonick put it:

> If the latest results do hold up, some of the most important questions in cosmology—how old the universe is, what it's made of and how it will end—will have been answered, only about 70 years after they were first posed. By the time the final chapter of cosmic history is written—further in the future than our minds can grasp—humanity, and perhaps even biology, will long since have vanished (157[25]:56).

The fact that *Time* magazine devoted an entire cover (and feature story to go with it) to the topic of "How the Universe Will End," is an inadvertent admission to something that evolutionists have long tried to avoid—the fact that the Universe had a beginning, and will have an ending. When one hears Sir James Jeans allude to the fact that "in some way matter which had not previously existed, came, or was brought, into being," the question that immediately comes to mind is: **Who** brought it into being?

What About the Big Bang?

Where are you right now? Are you sitting down with a cup of hot tea, ready to enjoy the few brief moments you can devote just to yourself? **Where** are you? Are you somewhere other than in your armchair at home? Or are you even at home? And if you are, in what city? In what state? In what country? And on what continent?

Astronomically speaking, you are on the third planet from the Sun, in a solar system of numerous other planets, only one of which—the one where you reside—sustains life. How? Why? These are intriguing questions worth pondering.

Throughout the whole of human history, people have contemplated not only their origin, but also their physical place in the Universe. The question of our ultimate origin weighs heavily on the human psyche. Science, to be sure, has brought its theories to bear on the subject. It is some of those theories that I would like to examine here.

Cosmology is the study of the Cosmos in all its aspects. The Cosmos, in simplest terms, is the space/mass/time Universe and all its arrays of complex systems. The cosmologist, whether under this title or not, has been around conceptually for centuries. Specifically, in the realm of science—as long as this term has been defined—we read about those of long ago such as Epicurus, Aristotle, and Copernicus, who sought answers to what they saw in the heavens. More recently in scientific history, we have people like Isaac Newton (1642-1727), Johannes Kepler (1571-1630), Willem de Sitter (1872-1934), Albert Einstein (1879-1955), Edwin Hubble (1889-1953), Georges Lemaître (1894-1966), Aleksandr Friedman (1889-1925), and George Gamow (1904-1968), each of whom made major contributions to understanding various theories and physical laws.

Nowadays, the scientific community includes numerous contributors of varying degrees. **Many viewpoints**, however, by no means implies **correct beliefs**. So, let us travel together down this road of cosmological descent—from the long-defunct Cartesian Hypothesis to modern versions of the Big Bang—and examine several of these theories in light of the scientific knowledge now available to us. As we proceed, let us heed the warning of the late cosmologist Sir Fred Hoyle (1915-2001), and his colleague, Chandra Wickramasinghe, in their book *Evolution from Space:* "**Be suspicious of a theory if more and more hypotheses are needed to support it as new facts become available, or as new considerations are brought to bear**" (1981, p. 135, emp. added).

The Evolution of a Theory

The science of cosmology, as we know it today, began, not surprisingly, with a look into the nearest and most readily observable astronomical environment—our solar system. Due to the sizable number of theories regarding the origin of our solar system, I will review only those that were of primary importance in the grand historical panorama.

The Cartesian Hypothesis, set down by the seventeenth-century French physician, mathematician, and philosopher René Descartes (1596-1650) in his *Principles of Philosophy,* postulated that our solar system had formed from a vast system of vortices running spontaneously. Out of these vortices, stars, comets, and planets emerged, each decaying into the next subsequent formation of matter, respectively. This particular conjecture did not sit well with some of Descartes' contemporaries, including Sir Isaac Newton, who made his disdain for Descartes' theory poignantly clear in a letter (penned on December 10, 1692) to evangelist Richard Bentley when he wrote: "The Cartesian hypothesis...can have no place in my system, and is plainly erroneous" (as quoted in Munitz, 1957, p. 212).

The next few hypotheses that flickered in history evolved their conceptual results from an initial rotating cloud of gas and/or dust known as a nebula. [Originally, the term "nebula" was applied to any distant object that appeared "fuzzy and extended" when viewed through a telescope; eventually, nebulae were identified as galaxies and star clusters.] Pierre S. Laplace (1749-1827), the distinguished French mathematician, presented his Nebular Hypothesis—a variation on the previously held hypotheses by Emanuel Swedenborg (1688-1772) and Immanuel Kant (1724-1804)—to the world in 1796. Laplace believed that, as the nebula rotated, it cooled and contracted, causing a discernible increase in rotational velocity, which eventually forced the matter that was located on the rim of the disc to overcome the gravitational attraction and be ejected from the cloud. The ejected matter then coalesced, forming a planet outside of the contracting nebula. This specific sequence of events continued until it formed a central portion of dense, rotating gases—what we know today as our Sun—and the outlying, orbiting planets (see Mulfinger, 1967, 4[2]:58). However, after failing a battery of mathematical and physical tests, these fanciful views ultimately were abandoned for the Planetesimal Hypothesis.

Heralded by T.C. Chamberlain (1843-1928) and F.R. Moulton (1872-1952), the Planetesimal Hypothesis started out with two initial stars, one of which was our Sun. The secondary star swept a near-collision path by the Sun, close enough to tear off two "arms" of matter on opposite sides. Over time, these arms coalesced to form planetesimals—tiny planets. This hypothesis followed in the footsteps of those that had preceded it (as well as a number of those yet to come) by failing to be scientifically accurate. Lyman Spitzer of Yale University demonstrated these failings: (1) the hot matter ripped from the Sun would not coalesce, but instead would continue to expand; and (2) one could not reconcile the angular momentum distribution of the solar system resulting from the interaction of the two passing stars (see Mulfinger, 4[2]:59-60).

The story of **modern** cosmology begins in the early parts of the twentieth century—a time when astronomers viewed the Universe as static, eternal, and limited in space to our own Milky Way Galaxy. Those views began to change in the early 1900s with the work of two American astronomers—Edwin Hubble and Vesto M. Slipher (1875-1969). Using one of the largest and most powerful telescopes available at the time, Hubble concluded that the Universe actually was much larger than just our galaxy. He determined that what were then known as "spiral nebulae," occurring millions of light-years away, were not part of the Milky Way at all, but rather were galaxies in their own right. [A light-year is the distance that light travels in a vacuum in one year—approximately 5.88 trillion miles. Distances expressed in light-years represent the time that light would take to cross that distance. For example, if an object were two million light-years away, it would require two million years, traveling at the speed of light, to traverse that distance.] Then, in 1929, Hubble reported a relationship between his distance information and some special analyses of light that had been carried out by Slipher (see Hubble, 1929).

Redshifts, Blueshifts, and Doppler Effects

In the decade spanning 1910-1920, Slipher (using a 24-inch, long-focus refractor telescope) had discovered the characteristic signature of atomic spectra in various far-flung galaxies. That discovery then led to another somewhat "unusual" finding. Examining a small sample of galaxies (which, at the time, were referred to as nebulae), he observed that the light frequencies those galaxies emitted were "shifted" toward the red portion of the spectrum (the concept of redshift is explained in detail below), which meant that they were receding from Earth. In 1913, Slipher reported the radial (or "line of sight") velocity of the Andromeda galaxy, and discovered that it was moving toward the Sun at a rate of 300 kilometers per second (see Slipher, 1913). This was taken as evidence in favor of the hypothesis that Andromeda was outside the Milky Way. [The Andromeda Galaxy is now considered a part of the "Local Group," which is an assortment of around thirty nearby galaxies (including the Milky Way) that is bound together gravitationally.] In 1914, Slipher reported radial velocities of 13 galaxies, and all but two were visualized as redshifts. By 1925, Slipher had compiled a list of 41 galaxies, and other astronomers had added four additional ones. Of the total of 45, 43 showed a redshift, which meant that only two were moving toward the Earth (see Gribbin, 1998, p. 76), while all the others were moving away from us.

These were, by all accounts, extraordinary observations. Using a far more sophisticated instrument (specifically, a larger, short-focus telescope that was better suited for this type of work), Edwin Hubble made the same types of discoveries in the late 1920s after Slipher had turned his attention to other projects. This "galactic redshift," Hubble believed, was an exceptionally stunning cosmic clue—a shard of evidence from far away and long ago. Why, Hubble wondered, should galactic light be shifted to the red, rather than the blue, portion of the spectrum? Why, in fact, should it be shifted at all?

From the very beginning, astronomers have attributed these shifts to what is known as the "Doppler effect." Named after Austrian physicist Christian Johann Doppler (1803-1853) who discovered the phenomenon in 1842, the Doppler effect refers to a specific change in the observed frequency of any wave that occurs when the source and the observer are in motion relative to each other; the frequency **increases** when the source and observer **approach** each other, and **decreases** when they **move apart**. By way of summary, the Doppler effect says simply that wavelengths grow longer (redshift) as an object recedes from the viewer; wavelengths grow shorter (blueshift) as an object approaches the viewer (see Figure 1 on the next page). [Color actually is immaterial in these terms, since the terms themselves apply to any electromagnetic radiation, whether visible or not. "Blue" light simply has a shorter wavelength than "red" light, so the use of the color-terms is deemed convenient.]

The light that we observe coming from stars is subject to the Doppler effect as well, which means that as we move toward a star, or as it moves toward us, the star's light will be shifted toward shorter (blue) wavelengths (viz., light that is emitted at a particular frequency is received by us at a higher frequency). As we move away from a star, or as it moves away from us, its light will be shifted toward longer (red) wavelengths (viz., light that is emitted at one frequency is received by us at a lower frequency). In theory then, a star's Doppler motion is a combination of both our motion through space (as the ob-

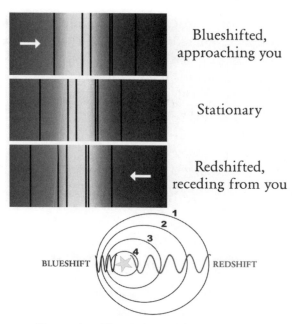

Blueshifted, approaching you

Stationary

Redshifted, receding from you

BLUESHIFT   REDSHIFT

Figure 1 – Blueshift/Redshift Depiction

server), and the star's motion (as we observe it). As it turns out, "the light from most galaxies exhibits a redshift roughly proportional to the galaxies' distance from us. Most cosmologists consider this pattern of redshifts to be evidence of cosmic expansion" (Repp, 2003, 39:270).

A word of caution is in order here. The Doppler effect, combined with the concepts of blueshift and redshift, can be somewhat confusing. It would be easy to assume that the expansion of the Universe is due solely to matter "flying through space" of its own accord. If that were true, then, of course, the Doppler effect would explain what is happening. But there is somewhat more to it than this. Cosmologists, astronomers, and astrophysicists suggest that the matter in the Universe is actually "at rest" with respect to the space around it. In other words, it is not the matter that is necessarily moving; rather, it is space itself that is doing the expanding. This means that, as space expands, whatever matter is present in that space simply gets "carried along for the ride." Thus, the particles of matter are not **really** moving apart on their own; instead, more space is appearing between the particles as the Universe expands, making the matter **appear** to move. Perhaps an illustration is appropriate here. [Bear with me; as you will see, the distinction that I am about to make has serious implications.]

More often than not, cosmologists use the example of a balloon to illustrate what they are trying to distinguish as "the true nature of the expanding Universe." Imagine, if you will, that someone has glued tiny shirt buttons to the surface of the balloon, and then commences to inflate it. As the balloon increases in size, the buttons will **appear to move** as they are carried along by the expansion of the balloon. But the buttons themselves are not actually moving. They are "at rest" on the balloon, yet are being "pushed outward" by the expansion of the medium around them (the latex of the balloon). Now, cosmologists suggest, compare this example to galaxies in space. The galaxies themselves can be "at rest" with respect to space, yet appear to be flying apart due to the expansion of the medium around them—space.

Almost all popular (and even most technical) publications advocate the view that the redshifts viewed in the expansion of the Universe are, in fact, attributable solely to the Doppler effect. But if it is true that the galaxies are actually at rest (although, admittedly, being "carried along" in an outward direction by the expansion of space itself, with its "embedded" galaxies), then the redshifts witnessed as a result of the expansion are not true Doppler shifts. To be technically correct, perhaps the galactic redshift should be called the "cosmological redshift." On occasion, when the "perceived motion" of the galaxies (as opposed to "real motion") is acknowledged at all, it sometimes is referred to as "Hubble flow." One of the few technical works with which I am familiar that acknowledges this fact (and even provides different formulae for the Doppler expansion versus the Hubble flow expansion) is *Gravitation*, by Misner, Thorne, and Wheeler (1973; see chapter 29).

Interestingly, as I was in the process of researching and writing this material, mathematician Andrew Repp of Hawaii authored a fascinating, up-to-date article on the nature of redshifts. In his discussion, Dr. Repp correctly noted that there are several known causes of redshifts (see Repp, 2003). One of the causes that he listed was the concept of "Hubble flow" expansion that I introduced above—which (again, interestingly) he labeled as "cosmological redshift" (39:271). As Repp remarked, this "expansion redshift" (a synonym for Hubble flow or cosmological redshift) "is caused by the expansion of space through which the wave is traveling, resulting in an 'expansion' (redshifting) of the wave itself.... [T]he expansion redshift would be the result of the motion of space itself." Yes, it would—which is exactly the point I was making in the above paragraphs. And, as Repp went on to acknowledge concerning expansion redshift: "It is the commonly accepted explanation for the redshifts of the distant galaxies" (39:271). Yes, it is.

But that is not quite the end of the story. There is evidence to support the idea that the galaxies themselves may, in fact, **actually be moving**, rather than simply being "at rest" while being carried along by the expansion of space. The Andromeda Galaxy (known as M31), which is among our nearest neighboring galaxies, presents a light spectrum that is blueshifted. If the Universe is expanding, how could that be? Apparently, the **Doppler motion** is large enough blueward to negate the **cosmological redshift expansion**, thereby allowing us to view a galaxy that has a blueshift. The implication of this is that the galaxy itself must be moving.

What could be responsible for that? Some astronomers have suggested that such movement may be attributable to the localized forces of gravity. Galaxies are known to clump together into clusters that can contain anywhere from a few dozen to a few thousand galaxies. [Clusters of clusters are known as "superclusters."] What holds these structures together? Presumably, it is gravity. That would imply that the objects composing the structures have orbits—which produce motion that are indeed Doppler in nature.

Andrew Repp expounded upon the concept I am discussing here, under the title of "gravitational redshift" in his article reviewing the various causes of redshifts, and specifically mentioned that "the expansion redshift differs from the gravitational redshift" (39:272). Yes, it does. As Dr. Repp commented, whereas the expansion redshift is the result of the motion of space itself, "gravitational redshift is the result of...the effects of gravity on spacetime" (39:271).

That being true, the light spectrum of any given galaxy will exhibit shifts that are the result of **both** the Doppler effect (due to actual motion) and the "cosmological redshift" (expansion redshift/Hubble flow—due to perceived motion). And how, exactly, would astronomers differentiate between the two? They wouldn't; observationally, there is no way to do so—which means that no one can say with accuracy how much of each exists. In fact, as Repp once again correctly noted, the Big Bang Model does not allow for "large-scale pattern of gravitational attraction, the mass distribution being assumed homogeneous; hence it predicts expansion redshifts but not (large-scale) gravitational redshifts" (39:272, parenthetical item in orig.). In point of fact, however, the commingling of cosmological redshift and gravitational redshift may well be one of the reasons that the calculation of the Hubble constant (discussed below) has been so problematic over the years. And this is why I stated earlier that the important distinction being discussed in this section has serious implications (different values for the Hubble constant result in varying ages for the Universe).

According to the standard Doppler-effect interpretation then, a redshifted galaxy is one that is traveling farther away from its neighbors. Hubble, and his colleague Milton Humason (1891-1972), plotted the distance of a given galaxy against the velocity with which it receded. By 1935, they had added another 150 points to the expansion data (see Gribbin, 1998, p. 81). They believed that the rate at which a galaxy is observed to recede is directly proportional to its distance from us; that is, the farther away a galaxy is from us, the faster it travels away from us. This became known as "Hubble's Law." Today, the idea that redshift is proportional to distance is a crucial part of distance measurement in modern astronomy. But that is not all. The concepts of (a) **an expanding Universe**, and (b) **the accuracy of redshift measurements**, form a critically important part of the foundation of modern Big Bang cosmology. As mathematician David Berlinski put it: "Hubble's law embodies a general hypothesis of Big Bang cosmology—namely, that the universe is expanding..." (1998, p. 34). One without the other is not possible. If one falls, both do. I will have more to say on this important point later.

Hubble and Humason's work gave cosmologists clues to the size of the Universe and the movement of objects within it. But while **astronomers** were peering through their telescopes at the Universe, **theoretical physicists** were describing that Universe in new ways. The first two models came from Albert Einstein and Willem de Sitter in 1917. Although they arrived at their models independently, both ideas were based on Einstein's General Theory of Relativity, and both scientists made adjustments to prevent expansion, even though expansion appeared a natural outcome of General Relativity. However, as knowledge about redshifts became more widespread, expansion was introduced as a matter of fact. [**Redshift** and **expansion** inevitably became the "twin pillars" upon which much of modern Big Bang cosmology was built. Interestingly, expansion itself also was built upon two pillars—**homogeneity** (matter is spread out uniformly) and **isotropy** (matter is spread out evenly in all directions). I will have more to say about all of this later, as well.] This was the case in 1922 with a set of solutions produced by Russian mathematician and physical scientist Aleksandr Friedman. Five years later, in 1927, the Belgian scholar Georges Lemaître produced a model incorporating a redshift-distance relation very close to that suggested by Hubble. If the Universe is expanding now, Lemaître calculated, then there must have been a time in the past when the Universe was in a state of contraction. It was in this state that the "primeval atom," as he called it, expanded to form atoms, stars, and galaxies. Lemaître had described, in its essential form, what is now known as the Big Bang, and scientists even today speak frequently of FL (Friedman-Lemaître) cosmology, which assumes the expansion of the Universe and its homogeneity (see Illingworth and Clark, 2000, p. 94).

The Big Bang Theory

While it was credited to Lemaître in his obituary, the eventual widespread acceptance of this hypothesis was due mainly to its leading constituent, Gamow. Even though it probably is not known widely today, the Big Bang–in its original "standard" form–actually came **before** the advent of the Steady State Theory and, ironically, was given its name (intended to be derogatory) by Hoyle as a result of a snide comment he made on a radio show for which he served as host (Fox, 2002, p. 65). In this section, I will discuss only the "standard" form of the Big Bang, leaving the discussion of the Big Bang's most recent variations for later.

In the beginning was the ylem...or so the theorists say. The "ylem"–an entirely hypothetical construct–was a primordial substance 10^{14} times the density of water, yet smaller in volume than a single proton. As one writer expressed it:

> Astonishingly, scientists now calculate that everything in this vast universe grew out of a region many billions of times smaller than a single proton, one of the atom's basic particles (Gore, 1983, 163:705).

The ylem (a.k.a. the "cosmic egg") was a "mind-bogglingly dense atom containing the entire Universe" (Fox, p. 69). [Where, exactly, the cosmic egg is supposed to have come from, no one knows; so far, no cosmic chicken has yet been sighted.] At some point in time, according to Big Bang theorists, the ylem reached its minimum contraction (at a temperature of 10^{32} Celsius–a 1 followed by 32 zeros!), and suddenly and violently expanded. Within an hour of this event, nucleosynthesis began to occur. That is to say, the light atoms we know today (e.g., hydrogen, helium, and lithium) had been manufactured in the intense heat. As the Universe expanded and cooled, the atoms started "clumping" together, and within a few hundred million years, the coalescing "clumps" began to form stars and galaxies (see Figure 2 on the next page). The heavier elements are assumed to have formed later via nuclear fusion within the cores of stars.

While the Steady State Theory had been widely accepted for more than a decade after its introduction, 1948 also was a good year for the competing Big Bang Theory. The first boost came from George Gamow and Ralph Alpher (currently, distinguished professor of physics, Union College, Schenectady, New York). They applied quantum physics to see how the Big Bang could make hydrogen and helium (plus minute amounts of lithium) —the elements thought to form 99% of the visible Universe— in a process called nucleosynthesis (see Gribbin, 1998, pp. 129-134). However, their theory was unable to account for elements heavier than helium; these would have to be made elsewhere. Geoffrey and Margaret Burbidge, Willy Fowler, and Fred Hoyle obliged—by suggesting that these other elements were manufactured in stars. To cap it all off, Fowler, Hoyle, and Robert Wagoner showed that the proportions of certain lighter-weight elements produced during the Big Bang matched almost exactly the proportions thought to exist in the solar system. This result, published in 1967, convinced many astronomers that the Big Bang was the correct description of the Universe's origin.

A decade later, the Big Bang was in full bloom. Robert Jastrow of NASA parroted the standard Big Bang refrain when he commented that, in the beginning, "all matter in the Universe was compressed into an infinitely dense and hot mass" that exploded. Then, over the many eons that followed, "the primordial cloud of the Universe expands and cools, stars are born and die, the sun and earth are formed, and life arises on the earth" (1977, pp. 2-3). With these statements, he was describing, of course, the essence of the Big Bang Theory, a concept that reigns supreme—in one form or another—as the current evolutionary explanation of the origin of the Universe. Berlinski assessed the theory's popularity as follows:

> As far as most physicists are concerned, the Big Bang
> is now a part of the structure of serene indubitability

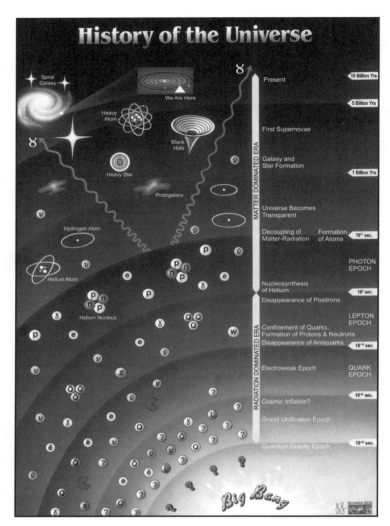

Figure 2 – Graphic representation of the alleged evolutionary origin of the Universe, from the Big Bang to the present, including the initial expansion phase, the production of matter, and galaxy formation. Courtesy of Center for European Nuclear Research (CERN), Geneva, Switzerland.

created by modern physics, an event undeniable as the volcanic explosion at Krakatoa. From time to time, it is true, the astrophysical journals report the failure of observation to confirm the grand design. It hardly matters. The physicists have not only persuaded themselves of the merits of Big Bang cosmology, they have persuaded everyone else as well (1998, p. 29).

Well, not quite everybody. It is true, of course, that cosmologists cling tightly to what they view as such a seemingly cohesive theory as the Big Bang. Princeton physicist Paul Steinhardt admitted:

An expanding universe, the microwave background radiation [discussed later—BT] and nucleosynthesis—these are the three key elements of the Big Bang model that seem to be very well verified observationally. They set a standard for any competing model (as quoted in Peterson, 1991, 139:232).

Truth be told, however, none of these concepts is without its own set of problems, and as a result, many scientists have acknowledged a number of critical flaws in the scenario you have just read. Hoyle stated the matter quite succinctly when he wrote:

As a result of all this, the main efforts of investigators have been in papering over holes in the big bang theory, to build up an idea that has become ever more complex and cumbersome. ...**I have little hesitation in saying that a sickly pall now hangs over the big bang theory**. When a pattern of facts becomes set against a theory, experience shows that the theory rarely recovers (1984, 92[5]:84, emp. added).

It is the view of many that the standard Big Bang not only **has not yet recovered**, but, in fact, **never will recover**. While that form of the Big Bang Theory has been in vogue throughout almost the whole of the scientific community, it nevertheless has fallen on hard times of late. [Revisions and variations of the Big Bang that still remain popular today will be discussed later.] As long ago as 1981, prominent astrophysicist Jayant Narlikar remarked:

These arguments should indicate to the uncommitted that the big-bang picture is not as soundly established, either theoretically or observationally, as it is usually claimed to be–astrophysicists of today who hold the view that "the ultimate cosmological problem" has been more or less solved may well be in for a few surprises before this century runs out (91:21).

Only two years later, evolutionist Don Page wrote: "There is no mechanism known as yet that would allow the Universe to begin in an arbitrary state and then evolve to its present highly ordered state" (1983, 304:40). Three years after that, renowned cosmologist John Gribbin reiterated the point when he wrote of the Big Bang Theory that "many cosmologists now feel that the shortcomings of the standard theory outweigh its usefulness..." (1986, 110[1511]:30). A decade-and-a-half later, one scientist, writing under the title of "The Bursting of the Big Bang," admitted that "while few people have seen the obituary **...the reality is that the immensely popular Big Bang Theory is dead**.... The Big Bang cannot explain the nature of the universe as we know it" (Lindsay, 2001, emp. in orig.). Berlinski, in "Was There a Big Bang?," wrote: "If the evidence in favor of Big Bang cosmology is more suspect than generally imagined, its defects are far stronger than generally credited" (1998, p. 37). Oh, how true. As it turns out, Narlikar, Page, Gribbin, and Lindsay were all correct. Scientists who advocated the Big Bang **were** in for "a few surprises." The standard Big Bang Theory **has** "outweighed its usefulness." And, yes, "the immensely popular Big Bang Theory **is** dead." Keep reading to find out why.

Scientific Reasons Why the Big Bang Theory Cannot be Correct

When one steps away from all the Big Bang propaganda, and carefully examines the foundation on which the concept itself rests, there is legitimate reason for concern. The theory, it appears, is haphazardly nestled on, and teeters on the brink of, some incredible assumptions–"incredible" in that each unstable assumption is built on top of another equally volatile supposition. It seems that, as this stack mounts, each subsequent assumption casts a shadow that hides from public view the visible uncertainties of the preceding one. Like an onion, as each layer is stripped back, it leaves only another lachrymose layer to be viewed. The time has come to peel back several of those layers, and expose what lies beneath. The Big Bang, as it turns out, is scientifically flawed.

An article ("The Self-Reproducing Inflationary Universe") by famed cosmologist Andrei Linde in the November 1994 issue of *Scientific American* revealed that the standard Big Bang Theory has been "scientifically brain dead" for quite some time. Linde (who, by the way, is the developer of two closely related variations of the Big Bang, known as the chaotic and the eternal inflationary models) is a professor of physics at Stanford University. He listed half a dozen extremely serious problems with the theory—problems that have been acknowledged for years (yet sadly, not always in a widely publicized fashion). Linde began his obituary for the Big Bang by asking the following question:

What Was There Before the Bang?

Scientists have been extremely successful, thus far, at diverting attention away from the obvious question: Where did the original material for the Big Bang come from? That is to say, what came before the Big Bang? John Gribbin voiced the opinion of many when he wrote: "The biggest problem with the Big Bang theory of the origin of the Universe is philosophical—perhaps even theological—**what was there before the bang**?" (1976, 259:15-16, emp. added). David Berlinski, writing in *Commentary* magazine, concluded:

> Such is the standard version of hot Big Bang cosmology—"hot" in contrast to scenarios in which the universe is cold, and "Big Bang" in contrast to various steady-state cosmologies in which nothing ever begins and nothing ever quite ends. **It may seem that this archeological scenario leaves unanswered the question of how the show started and merely describes the consequences of some Great Cause that it cannot specify and does not comprehend** (1998, p. 30, emp. added).

It's not just that "it may **seem**" that the Big Bang Theory "leaves unanswered the question of how the show started." It's that it **does** leave such questions unanswered! Linde admitted that there is a chicken-and-egg problem involved here. In his *Scientific American* article, he noted:

> The first, and main, problem is the very existence of the big bang. **One may wonder, What came before?** If space-time did not exist then, how could everything appear from nothing? What arose first: the universe or the laws governing it? Explaining this initial singularity—where and when it all began—still remains the most intractable problem of modern cosmology (1994, 271[5]:48, emp. added).

Yes, "one may wonder." But that is not all about which one may wonder, as Linde pointed out later when he inquired, "If there was no law, how did the Universe appear?" (as quoted in Overbye, 2001). British physicist Stephen Hawking asked:

> **What is it that breathes fire into the equations and makes a universe for them to describe?** The usual approach of science of constructing a mathematical model cannot answer the question of why there should be a universe for the model to describe.... Even if there is only one possible unified theory, it is just a set of rules and equations (1988, p. 174, emp. added).

In a chapter titled "Science and the Unknowable" in one of his books, humanist Martin Gardner followed Hawking's and Linde's lead:

> Imagine that physicists finally discover all the basic waves and their particles, and all the basic laws, and unite everything in one equation. We can then ask, "Why **that** equation?" It is fashionable now to conjecture that the big bang was caused by a random quantum fluctuation in a vacuum devoid of space and time. But of course such a vacuum is a far cry from nothing. **There had to be quantum laws to fluctuate. And why are there quantum laws?...There is no escape from the superultimate questions: Why is there something rather than nothing, and why is the something structured the way it is?** (2000, p. 303, emp. added).

- 50 -

British cosmologist John Barrow addressed the issue in a similar fashion when he wrote:

> At first, the absence of a beginning appears to be an advantage to the scientific approach. There are no awkward starting conditions to deduce or explain. **But this is an illusion. We still have to explain why the Universe took on particular properties**—its rate expansion, density, and so forth—at an infinite time in the past (2000, p. 296, emp. added).

Gardner and Barrow are correct. And science, as impressive as it is, cannot provide the solutions to such problems. Nancey Murphy and George Ellis discussed this very point in their book, *On the Moral Nature of the Universe:*

> Hence, we note the fundamental major metaphysical issues that purely scientific cosmology by itself cannot tackle—the problem of existence (what is the ultimate origin of physical reality?) and the origin and determination of the specific nature of physical laws—for these all lie outside the domain of scientific investigation. The basic reason is that there is no way that any of these issues can be addressed experimentally. The experimental method can be used to test existing physical laws but not to examine why those laws are in existence. One can investigate these issues using the hypothetico-deductive method, but one cannot then conduct physical, chemical, or biological experiments or observations that will confirm or disconfirm the proposed hypotheses (1996, p. 61).

Entire Universes from Black Holes?

In the opinion of cosmologist Hannes Alfven, the ylem never could have attained the incredible density postulated by the Big Bang Theory (see Mulfinger, 1967, 4[2]:63). But what if it had? Astronomer Paul Steidl offered yet another puzzle.

> If the universe is such and such a size now, they argue, then it must have been smaller in the past, since it is observed to be expanding. If we follow this far enough backward in time, the universe must have been very small, as small as we wish to make it by going back far enough. This leads to all sorts of problems which would not even come up if scientists were to realize that time

can be pushed back only so far; they do not have an infinite amount of time to play with.... To bring all the matter in the universe back to the same point requires 10 to 20 billion years. Astronomers postulate that at that time all the matter in the universe was at that one spot, and some explosion of unimaginable force blew it apart at near light-speeds. What was that matter like, and how did it get there in the first place? And how did it come to be distributed as it is now? These are the basic questions that cosmological models try to answer, but the solutions continue to be elusive. With the entire universe the size of a pinpoint,* normal physical laws as we know them must have been drastically different. There is no way scientists can determine what conditions would have been like under these circumstances. One could not even tell matter from energy. Yet astronomers continue to make confident assertions about just what went on during the first billionth of a second! (1979, p. 195).

Interestingly, at the place in Steidl's quote where you see the asterisk ("...with the universe the size of a pinpoint*..."), there was a corresponding asterisk at the bottom of the page, indicating a footnote that included this statement: "Question: **Why did the universe not become a black hole?**" (emp. added). Good question. As Gerardus Bouw wrote in an article titled "Cosmic Space and Time": "In order to save the Big Bang cosmology, are we to believe that the...physics of black holes does not work for the universe?" (1982, 19[1]:31). If all the matter/energy in the Universe were packed into a point "many billions of times smaller than a single proton," why would that not constitute a black hole? [NOTE: The reader who is interested in investigating further the concept of black holes (including whether or not they actually exist) may wish to read: (a) Hazel Muir's article, "Death Star," in the January 19, 2002 issue of *New Scientist;* and (b) "New Theories Dispute the Existence of Black Holes," (2002).]

Interestingly, some scientists actually have now begun to suggest that the Universe **did** evolve from a black hole. Lee Smolin, professor of physics at Pennsylvania State University, suggested exactly that in his book, *The Life of the Cosmos: A New View of Cosmology, Particle Physics, and the Meaning of Quantum Physics* (1995). In a chapter titled "The Theory of the Whole Universe" that he authored for John Brockman's book, *The Third Culture*, Dr. Smolin discussed his view of what he refers to as "cosmological natural selection."

> It seemed to me that the only principle powerful enough to explain the high degree of organization of our universe—compared to a universe with the particles and forces chosen randomly—was natural selection itself. The question then became: Could there be any mechanism by which natural selection could work on the scale of the whole universe?

> Once I asked the question, the answer appeared very quickly: the properties of the particles and the forces are selected to maximize the number of black holes the universe produces.... [A] new region of the universe begins to expand **as if from a big bang, there inside the black hole**.... I had a mechanism by which natural selection would act to produce universes with whatever choice of parameters would lead to the most production of black holes, since a black hole is the means by which a universe reproduces—that is, spawns another (1995, p. 293, emp. added).

Immediately following Smolin's chapter in *The Third Culture*, cosmologist Sir Martin Rees (Britain's Astronomer Royal) offered the following invited response:

> Smolin speculates—as others, like Alan Guth, have also done—that inside a black hole it's possible for a small region to, as it were, sprout into a new universe. We don't see it, but it inflates into some new dimension.... What that would mean is that universes which can therefore produce lots of black holes, would have more progeny, because each black hole can then lead a new universe; whereas a universe that didn't allow stars and black holes to form would have no progeny. Therefore Smolin claims that the ensemble of universes may evolve not randomly but by some Darwinian selection, in favor of the potentially complex universes.

> My first response is that we have no idea about the physics at these extreme densities, so we have no idea

whether the physics of the daughter universe would resemble that of the parent universe. But one nice thing about Smolin's idea, which I don't think he realized himself in his first paper, is that it's in principle testable....

The bad news is that I don't see any reason to believe that our universe has the property that it forms more black holes than any other slightly different universe. There are ways of changing the laws of physics to get **more** black holes, so in my view there are arguments **against** Smolin's hypothesis. It's just everyday physics, or **fairly** everyday physics, that determines how stars evolve and whether black holes form and I can tell Smolin that our universe doesn't have the properties that maximize the chance of black holes. I could imagine a slightly different universe that would be even better at forming black holes. If Smolin is right, then why shouldn't our universe be like that? (as quoted in Smolin, 1995, pp. 298,299, emp. in orig.).

The essence of Sir Martin's question—"If Smolin is right, why shouldn't our universe be like that?"—applies to more than just Dr. Smolin's particular theory. It applies across the board to any number of theories: "If ____ is right, why shouldn't our universe be like ____?" Which is exactly one of the points I am trying to get across. The simple fact is, in many of these "off-the-wall" theories, the Universe **is not** "like that." In commenting on Smolin's ideas, Berlinski wrote:

> **There is, needless to say, no evidence whatsoever in favor of this preposterous theory**. The universes that are bubbling up are unobservable. So, too, are the universes that have been bubbled up and those that will bubble up in the future. Smolin's theories cannot be confirmed by experience. Or by anything else. What law of nature could reveal that the laws of nature are contingent?
>
> **Contemporary cosmologists feel free to say anything that pops into their heads**. Unhappy examples are everywhere: absurd schemes to model time on the basis of the complex numbers, as in Stephen Hawking's *A Brief History of Time*; bizarre and ugly contraptions for cosmic inflation; universes multiplying

beyond the reach of observation; white holes, black holes, worm holes, and naked singularities; theories of every stripe and variety, all of them uncorrected by any criticism beyond the trivial. The physicists carry on endlessly because they can (1998, p. 38, emp. added).

"Carrying on endlessly," unfortunately, has not helped matters. Once again, keep reading.

Redshift and Expansion Problems

As I mentioned earlier, the twin ideas of (a) the **accuracy of redshift measurements** and (b) **an expanding Universe** form a critically important part of the foundation of modern Big Bang cosmology. As late as 1979, scientists were shocked to learn that two of the methods that had been used to derive many of their measurements regarding ages and distances within the Universe—the Hubble constant (see next paragraph) and redshift measurements (to be discussed shortly)—were in error.

The value of the Hubble constant (H_0—the constant of proportion between relative velocity and distance that is used to calculate the expansion rate of the Universe) is expressed in kilometers per second per megaparsec [one parsec equals just a little over 3 light-years (3.2616 to be exact); a megaparsec (Mpc) is one million parsecs]. Initially, the Hubble constant was set by Hubble himself at around 500 km/sec/Mpc (Hubble, 1929). Since then, it has been revised repeatedly. In fact, of late, astronomical theory has run headlong into a series of nasty problems regarding the continued recalibration of the so-called Hubble **constant**. Observe, for example, the data in Table 1 on the next page (adapted from DeYoung, 1995).

In an article he wrote on "The Hubble Law," physicist Don DeYoung noted:

The Hubble constant cannot be measured exactly, like the speed of light or the mass of an electron. Aside from questions about its possible variation in the past, there is simply no consensus on its value today....

AUTHOR	PUBLICATION YEAR	HUBBLE CONSTANT	UNIVERSE AGE (billions of years)
Hubble	1929	500*	2
Harwit	1973 (p. 61)	75	9
Pasachoff	1992 (p. 366)	36	18
Gribbin	1993	26	25
Freedman	1994	65-99	8-12
Hawking	1994 (p. 46)	43	15
Kuhn	1994 (p. 556)	54	12
Matthews	1994	80	8
Ross	1994 (p. 95)	38	17
Schmidt	1994	64-82	10-12
Wolff	1994 (p. 164)	50	13
MacRobert	2003 (pp. 16-17)	71	13.7

Table 1 – Hubble constant values, 1929-2003. *The original value of the Hubble constant was not well defined because of scatter in the data (see Gribbin, 1998, p. 79, figure 4.1A). Estimates range from 320 to 600 km/sec/Mpc, but perhaps the most popular views sets Hubble's initial estimate at around 500 km/sec/Mpc.

Today there are two popular competing values for the Hubble constant. A smaller value of about $H = 50$ is promoted by Allan Sandage, Gustav Tammann and colleagues. This constant results in a universe age of about 19.3 billion years. A larger value, $H = 100$, is preferred by many other astronomers: Gerard de Vaucouleurs, Richard Fisher, Roberta Humphreys, Wendy Freedman, Barry Madore, Brent Tully and others. The $H = 100$ value gives a universe age half that of Sandage, "just" 9 billion years or less, depending on the gravity factor used (1995, 9[1]:9, emp. added).

DeYoung was correct when he suggested in regard to the Hubble constant that "there is simply no consensus on its value today." Gribbin, in his book, *In Search of the Big Bang*, remarked concerning the disagreement between the two camps specifically mentioned by DeYoung (Sandage, et al., and Vaucouleurs, et al.): "Neither seems willing to budge" (1998, p. 188). Little wonder. As Gribbin also observed: **"Hubble's constant is the key number in all of cosmology**. Armed with an accurate value of *H* and redshift measurements, it would be possible to calculate the distance to any galaxy" (pp. 187-188, emp. added).

But "an accurate value of *H*" has thus far eluded astronomers, cosmologists, and physicists. Based on measurements of 20 Cepheid variable stars from the Virgo Cluster of galaxies, the Hubble constant has been measured at 80 km/sec/Mpc (see Freedman, et al., 1994; Jacoby, 1994). [Assuming that the Big Bang theory for the origin of the Universe is correct, that would correspond to an age of the Universe of about 8 billion years.] Yet, as DeYoung pointed out, another group of astronomers, led by Allan Sandage, has claimed that the Hubble constant should be set at about 50 km/sec/Mpc (see Cowen, 1994), which (depending on the application of various correction factors) would make the Universe somewhere in the range of 13-20 billion years old (Travis, 1994).

Still another group of astronomers has argued that astronomical theories would require a Hubble constant of 30 km/sec/Mpc (Bartlett, et al., 1995). As of this writing, according to data from NASA's Wilkinson Microwave Anisotropy Probe [WMAP] (as reported in an article, "Turning a Corner on the New Cosmology," in the May 2003 issue of *Sky and Telescope*), the latest value for the Hubble constant has been set at 71 +/- 4 km/sec/Mpc, yielding an age for the Universe of 13.7 billion years (see MacRobert, 105[5]:16-17). Well-known astronomer Halton Arp (discussed below) has referred to what he calls the continuing "soap opera of conflicting claims about the value of the Hubble constant" (1999, p. 234), and commented that numerous "corrections" frequently are required to make the available data "fit" (p. 153).

Christopher DePree and Alan Axelrod admitted: "Actually the precise value of H_0 is the subject of dispute" (2001, p. 328). That is a mild understatement, since the current value of the Hubble constant varies between 50 and 75 km/sec/Mpc (see

Cowen, 1994; Illingworth and Clark, 2000, p. 198). [It is important to understand that the value of the Hubble "constant" is not a trivial matter. As DePree and Axelrod went on to note: "A different Hubble constant gives the universe a different age" (p. 328). This fact is clearly evident from the data in Table 1 on page 56.]

In the minds of some, one of the most significant problems facing Big Bang cosmology today has to do with the concept of **redshift**. Perhaps the easiest way to understand redshift is to imagine the sound coming from a siren on a fire engine. Once that fire engine passes, the pitch drops. The siren does not actually change pitch; rather, the sound waves of an **approaching** fire engine are made **shorter** by the approach of the sound source, where the waves of the **departing** fire engine are made **longer** by the receding of the sound source (see Figure 1). Light (or electromagnetic radiation) from stars or galaxies behaves in exactly the same manner. As noted earlier, an approaching source of light or radiation emits shorter waves (relative to an observer). A receding source emits longer waves (again, relative to the observer). Thus, the radiation or light of a source moving **toward** an observer will be "shifted" toward the **blue** end of the wavelength scale. The radiation or light of a source moving **away** from the observer "shifts" toward the **red** end of the light spectrum. The amount of shift is a function of the relative speed. A body approaching or receding at a high speed will show a greater shift than one approaching or receding at a low speed.

Illingworth and Clark observed in regard to the Hubble constant: "The velocity can be measured accurately from the redshift in the galaxy's spectrum" (2000, p. 198). But what if the redshift measurements themselves are incorrect? That, by definition, would affect the Hubble constant, which in turn would alter the size and age estimates of the Universe, which in turn would impact cosmic evolution, etc.

The redshift controversy has been elucidated most effectively by American astrophysicist Halton Arp, currently at the Max Planck Institute for Astrophysics in Munich, Germany.

Arp—who has been called "the world's most controversial astronomer" (Kaufmann, 1982)—has suggested that redshifts are not necessarily attributable to the Doppler effect (see Amato, 1986; Bird, 1987, pp. 5,8). Dr. Arp is difficult to dismiss; he worked with Edwin Hubble himself, and formerly was at the Mt. Palomar Observatory. He has studied the relationship between quasars (see definition below) and what he refers to as "irregular" galaxies, and, on the basis of his observations, has opposed the standard belief in the correlating relationship between an object's redshift and its velocity. In fact, Arp has found what he calls "enigmatic and disturbing cases," where two apparently connected objects that seem to be the same distance away, actually have significantly different redshift values (see Sagan, 1980, p. 255; Arp, 1987; Cowen, 1990a; Arp, 1999).

For example, by taking photographs through the big telescopes, Arp discovered that many pairs of quasars that have extremely high redshift values (and therefore are thought to be receding from us very rapidly—which means that they must be located at a great distance from us) are associated physically with galaxies that have low redshifts, and thus are thought to be relatively close. Dr. Arp has produced extremely impressive photographs of many pairs of high-redshift quasars that are located symmetrically on either side of what he proposes are their parent, low-redshift galaxies [see "Arp's Anomalies" in Appendix B]. Such pairings, Arp suggests, occur far more frequently than the probabilities of random placement should allow. Mainstream astrophysicists have tried to explain away Arp's observations of connected galaxies and quasars as being "illusions" or "coincidences of apparent location." But, the large number of physically associated quasars and low-redshift galaxies that he has photographed and cataloged defies such an explanation. It simply happens too often. As Arp himself commented: "One point at which our magicians attempt their sleight-of-hand is when they slide quickly from the Hubble, **redshift-distance** relation to **redshift velocity** of expansion" (as quoted in Martin, 1999, p. 217, emp. added). In his volume, *Seeing Red: Redshifts, Cosmology and Academic Science*, Arp wrote:

But if the cause of these redshifts is misunderstood, then distances can be wrong by factors of 10 to 100, and luminosities and masses will be wrong by factors up to 10,000. **We would have a totally erroneous picture of extragalactic space, and be faced with one of the most embarrassing boondoggles in our intellectual history** (1999, p. 1, emp. added).

All of this means, of course, that the redshift may be virtually useless for calculating the recession speed of distant galaxies, and would completely destroy one of the main pillars of the expanding-Universe idea. Meteorologist Michael Oard noted:

What if the redshift of starlight is unrelated to the Doppler effect, i.e., the principle that relative motion changes the observed frequency of the light emitted from a light source? Many of the deductions of mainstream cosmology would fold catastrophically (2000, 14[3]:39).

Astronomer William Kaufmann concluded in an article he wrote about Arp titled "The Most Feared Astronomer on Earth":

If Arp is correct [about redshifts not being distance indicators—BT], if his observations are confirmed, he will have single-handedly shaken all modern astronomy to its very foundations. **If he is right, one of the pillars of modern astronomy and cosmology will come crashing down in a turmoil unparalleled since Copernicus dared to suggest that the sun, not the earth, was at the center of the solar system** (1981, 89[6]:78, emp. added).

Or, as Fox lamented:

Redshifts are not, in and of themselves, a sign of a star's age or distance, and yet redshifts have become intrinsically entwined with how we determine not just the speed of any given object, but also how old and how far away it is. **If the interpretation of redshift is wrong, then all the proof that the universe is expanding will disappear. It would undermine everything that's been mapped out about the heavens**. Not only would the big bang theory come crashing down, but scientists wouldn't be able to determine how the nearest galaxy is moving, much less how the whole universe behaves (2002, p. 129, emp. added).

What is going on here? The history of this fascinating story actually harks back to the 1940s. But Arp's work has updated it considerably. Berlinski has told the tale well.

> At the end of World War II, astronomers discovered places in the sky where charged particles moving in a magnetic field sent out strong signals in the radio portion of the spectrum. Twenty years later, Alan Sandage and Thomas Mathews identified the source of such signals with optically discernible points in space. These are the quasars–**quasi stellar radio sources**.

> Quasars have played a singular role in astrophysics. In the mid-1960's, Maarten Schmidt discovered that their spectral lines were shifted massively to the red. If Hubble's law were correct, quasars should be impossibly far away, hurtling themselves into oblivion at the far edge of space and time. But for more than a decade, the American astronomer Halton Arp has drawn the attention of the astronomical community to places in the sky where the expected relationship between redshift and distance simply fails. Embarrassingly enough, many quasars seem bound to nearby galaxies. The results are in plain sight: there on the photographic plate is the smudged record of a galaxy, and there next to it is a quasar, the points of light lined up and looking for all the world as if they were equally luminous.

> These observations do not comport with standard Big Bang cosmology. If quasars have very large redshifts, they must (according to Hubble's law) be very far away; if they **seem** nearby, then either they must be fantastically luminous or their redshift has not been derived from their velocity.... But whatever the excuses, a great many cosmologists recognize that quasars mark a point where the otherwise silky surface of cosmological evidence encounters a snag (1998, pp. 32-33, parenthetical item and emp. in orig.).

That "snag" is what Halton Arp's work is all about. [See Appendix B for additional information concerning Dr. Arp's data and conclusions.] Compounding the problem related to the quasars is the concept of what might be termed "premature aging." Cosmologists now place the Big Bang event at 13.7 billion years ago (see Brumfiel, 2003, 422:109; Lemonick, 2003,

161:45), **and the specific beginnings of galaxy formation somewhere between 800,000 to 1,000,000 years after that** (Cowen, 2003, 163:139). Hence, radiation coming from an object 13 billion light-years away supposedly began its journey approximately a billion years after the Big Bang, when the object was somewhat less than a billion years old. Such distant objects should show relatively few signs of development, but observations within the last decade have threatened such concepts. For example, the Röentgen Satellite found giant clusters of quasars more than 12 billion light-years away (Cowen, 1991a), and astronomers have detected individual quasars at 12-13 billion light-years away (Cowen, 1991b; 2003).

The problem is that quasars–those very bright, super-energetic star-like objects–are thought to have formed **after** their hypothetical energy sources and resident galaxies had emerged. Hence, very distant quasars and quasar clusters represent **too much organization too early in the history of the Universe**. This is indeed problematic. As one scientist put it, the Big Bang theorist suddenly "finds himself in the position of a cement supplier who arrives after the house is already built" (Major, 1991b, 11:23).

In the January 31, 1997 issue of *Science*, Hans-Dieter Radecke wrote that modern cosmology's dependence on "interpretations of interpretations of observations" makes it essential that "we should not fall victim to cosmological hubris, but stay open for any surprise" (275:603). Good advice, to be sure. And a mere six years after he made that comment, those "surprises" began. The March 1, 2003 issue of *Science News* reported several "surprises" that "do not comport with standard Big Bang cosmology" (to use Berlinski's words). First, astronomical research indicates that

> a surprising number of galaxies grew up in a hurry, appearing old and massive even when the universe was still very young. If this portrait of precocious galaxies is confirmed by larger studies, astronomers may have to revise the accepted view of galaxy formation.... In mid-December [2002], scientists announced in a press release that they had found a group of distant

galaxies that were already senior citizens, chock-ablock with elderly, red stars a mere 2 billion years after the Big Bang. The same team found another surprise. **Some of those galaxies were nearly as large as the largest galaxies in the universe toda**y (Cowen, 2003, 163:139, emp. added).

Talk about "premature aging"!

Second, on January 7, 2003, another team of scientists reported that it had found "the oldest, and therefore most distant, galaxy known. If confirmed, the study indicates that some galaxies were in place and forming stars at a prolific rate when the universe, now 13.7 billions years old, was just an 800-million-year-old whippersnapper" (Cowen, 163:139).

Third,

> at a galaxy-formation meeting in mid-January [2003] in Aspen, Colorado, [Richard] Ellis [of the California Institute of Technology in Pasadena] reported other evidence that the 2-billion-year-old universe was populated with as many galaxies marked by red, senior stars as by blue, more youthful stars.... **If accurate, this new view of galactic demography might force astronomers to rethink the fundamentals of galaxy formation** (Cowen, 163:140, emp. added).

Talk about "cosmological evidence encountering a snag"! What an understatement. A number of astronomers, of course, have preferred to simply ignore work like Arp's, which "does not comport" with standard Big Bang cosmology. "Others," wrote Berlinski, "have scrupled at Arp's statistics. Still others have claimed that his samples are too small, although they have claimed this for every sample presented and will no doubt continue to claim this when the samples number in the billions" (p. 33). Sadly, because Arp's views do not come anywhere close to supporting the status quo, he even has been denied telescope time for pursuing this line of research (see Gribbin, 1987, Marshall, 1990). [As William Corliss commented (somewhat sarcastically) in discussing this issue: "Some astronomers, according to news items in scientific publications, have heard enough about discordant redshifts and would rather see scarce telescope time used for other types of work" (1983).] If Dr. Arp is correct, however (and there is compelling evidence to indicate that he is—see next paragraph), then the Universe is not acting in a way that is consistent with the Big Bang Theory.

Support for Arp's conclusions arrived in the form of research performed by another American—I.E. Segal—a distinguished mathematician who also happens to be one of the creators of modern function theory, and who is a member of the National Academy of Sciences. He and his coworkers studied the evidence for the recessional velocities of galaxies over the course of a twenty-year period. The experimental results of their research, as it turns out, were quite disturbing to Big Bang theorists, because those results are sharply at odds with predictions made by Big Bang cosmology.

Our place in the Universe. This composite radio light image (as seen in visible light) illustrates the enigmatic "high-velocity clouds" of gas (depicted by the various colors) above and below the plane of the Milky Way Galaxy (seen in white). Photo courtesy of NASA.

Galaxies, as everyone involved in cosmology readily acknowledges, are critical when it comes to verification (or non-verification, as the case may be) of Hubble's law, because it is by observing galaxies that the crucial observational evidence for the Big Bang must be uncovered. When Segal examined redshift values within various galaxies during his two-decade-long study,

[t]he linear relationship that Hubble saw, Segal and his collaborators cannot see and have not found. Rather, the relationship between redshift and flux or apparent brightness that they have studied in a large number of complete samples satisfies a quadratic law, the redshift varying as the square of apparent brightness (Berlinski, 1998, pp. 33-34).

Segal concluded: "By normal standards of scientific due process, the results of [Big Bang] cosmology are illusory." He then went on to claim that Big Bang cosmology

owes its acceptance as a physical principle primarily to the uncritical and premature representation [of the redshift-distance relationship–BT] as an empirical fact. …Observed discrepancies…have been resolved by a pyramid of exculpatory assumptions, which are inherently incapable of noncircular substantiation (as quoted in Berlinski, p. 33).

More than one cosmologist has dismissed Segal's claims (which, remember, are based on twenty-years' worth of scientific research) with what Berlinski called "a great snort of indignation." But, observed Berlinski, "the discrepancy from Big Bang cosmology that they reveal is hardly trivial" (p. 34).

Indeed, the discrepancy is "hardly trivial." As I noted earlier, the idea that the Universe is expanding is listed as one of the three main support pillars for Big Bang cosmology (see Fox, pp. 56,120). Both the **fact** of expansion, and the **rate** of expansion, have as part of their foundation the redshift values of stellar objects (specifically, galaxies)–redshift values that now are being called into question in a most rigorous manner by distinguished astronomers and mathematicians. Surely, it is evident that a serious re-evaluation of these matters is in order. Fox stated the relationship well when she wrote:

Many…people strike at the very heart of the big bang theory: expansion. While, as mentioned earlier, an expanding universe doesn't require that the universe began with a bang, the big bang theory certainly requires an expanding universe. **If it turns out that galaxies and stars aren't receding from each other, then the entire theory would fall apart** (p. 126, emp. added).

Yes, it certainly would. But it gets worse. In his critique of the standard Big Bang Theory in *Scientific American*, Andrei Linde listed as number four in his list of six "highly suspicious underlying assumptions" (as he called them)—"the expansion problem."

> The fourth problem deals with the timing of the expansion. In its standard form, the big bang theory assumes that all parts of the universe began expanding simultaneously. **But how could the different parts of the universe synchronize the beginning of their expansion? Who gave the command?** (1994, 271[5]: 49, emp. added).

Who indeed? George Lemaître, who originally postulated the idea of the Big Bang, suggested that the Universe started out in a highly contracted state and initially expanded at a rapid rate. The expansion slowed down and ultimately came to a halt, during which time, galaxies formed and gave rise to a new expansion phase that then continued indefinitely. One of the difficulties here is that the Universe is supposed to be all there is. That is to say, it is self-contained. [The late astronomer of Cornell University, Carl Sagan, opened his television extravaganza *Cosmos* (and his book by the same name) with these words: "The Cosmos is all that is or ever was or ever will be" (1980, p. 4). That is about as good a definition of a "self-contained" Universe as one could hope to find.]

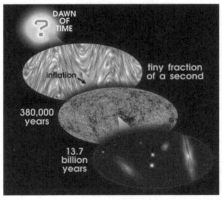

Figure 3 – Artist's concept of crucial periods in the development of the Universe according to Big Bang inflationary cosmology

But, "somehow," the expansion conveniently started moving again, after the galaxies had time to form in a non-moving, static Universe. According to Newton's first law of motion, however, an object will continue in whatever state of motion it is in, unless acted upon by an unbalanced external force. In other words, if it were sitting still, it would have to remain like that (meaning—no further expansion!). But in the Big Bang, the Universe just "picks up" and continues to expand after the galaxies finally get formed. Sir Fred Hoyle, addressing this very point, put it succinctly when he referred to the Big Bang model as a

> dull-as-ditchwater expansion which degrades itself adiabatically [without loss or gain of heat—BT] until it is incapable of doing anything at all. The notion that galaxies form, to be followed by an active astronomical history, is an illusion. Nothing forms; the thing is dead as a doornail (1981, 92:523).

Ouch!

The idea of a "brief hiatus" of sorts for galaxy formation is one of those *ad hoc*, quickly improvised hypotheses that had to be added to keep the Big Bang Theory alive. There certainly is no physical basis for it—which was what Dr. Hoyle's "dull as ditchwater" comment was intended to reflect. A "bang" does not allow for starts and stops. Once a bomb goes off, an observer hardly expects gravitation to cause the shrapnel to come back together and form clumps, no matter how near (or far) the pieces travel from the location of the initial explosion.

Cosmic Microwave Background Radiation

In 1978, Arno Penzias and Robert Wilson were honored with the Nobel Prize in physics for their discovery of the cosmic microwave background radiation (referred to variously in the literature as CMB, CMR, or CBR; I will use the CMB designation throughout this discussion). The two researchers from Bell Laboratory serendipitously stumbled onto this phenomenon in June 1964, after first thinking it was an equipment malfunction. For a short while, they even attributed the background noise to what they referred to as "white dielectric material"—i.e., bird droppings (Fox, 2002, p. 78). The electromagnetic radiation they were experiencing was independent of the spot in the sky where they were focusing the antenna, and was only a faint "hiss" or "hum" in its magnitude. The microwaves, which can be related to temperature, produced the equivalent of approximately 3.5 K background radiation at 7.3 cm wavelength ("K" stands for Kelvin, the standard scientific temperature scale; 0 K equals absolute zero—the theoretical point at which all motion ceases: -459° Fahrenheit or -273° Celsius). Unable to decide why they were encountering this phenomenon, Penzias and Wilson sought the assistance of Robert Dicke at Princeton University who, with his colleagues, immediately latched onto this noise as the "echo" of the Big Bang. A prediction had been made prior to the discovery, that if the Big Bang were true, there should be some sort of constant radiation in space, although the prediction was for a temperature several times higher (see Weinberg, 1977, p. 50; Hoyle, et al., 2000, p. 80).

Previously, in the section on the Steady State Theory, I referred to the fact that a "new theoretical concept" eventually would be responsible for dethroning that theory. That reference was to Penzias and Wilson's discovery of the existence of the cosmic microwave background radiation. Described by some evolutionists as the "remnant afterglow of the Big Bang," it is viewed as a faint light shining back to the beginning of the Universe (well, close to the beginning...say, within 300,000 to 400,000 years or so). This radiation, found in the form of microwaves, has been seized upon by proponents of the Big Bang Theory as proof of an initial catastrophic beginning—the "bang" —of our Universe. However, the temperature estimates of space were first published in 1896, even prior to George Gamow's birth in 1904 (see Guillaume, 1896). C.E. Guillaume's estimation was 5-6 K, and rather than blaming that temperature on some type of "Big Bang" explosion, he credited the stars belonging to our own galaxy.

The cosmic background radiation spelled almost instant doom for the Steady State Theory, because the theory did not predict a background radiation (since there was no initial outpouring of radiation in that model). Plus, there was no way to introduce the idea of such background radiation into the existing theory. Therefore, the Quasi-Steady-State Theory, a slight variation by Hoyle, Burbidge, and Narlikar, was formed to try to make sense of this "chink" in the armor of the Steady State Theory. The British science journal *Nature* stated it well: "Nobody should be surprised, therefore, if the handful of those who reject the Big Bang claim the new data as support for their theories also" (see "Big Bang Brouhaha," 1992, 356:731). The prediction made by *Nature* was right on target. The CMB radiation data have indeed been used by almost all theorists as an *ad hoc* support for their views. A logical question to ask would be: "Do these various groups all claim it on the same scientific grounds?" The answer, of course, is no.

Speaking of the CMB radiation, Joseph Silk referred to the results as "the cornerstone of Big Bang cosmology" (1992, p. 741). There can be no doubt that there exists a cosmic electromagnetic radiation on the microwave order, and that its temperature correlation is approximately 3 K (technically 2.728 K; see Harrison, 2000, p. 394). This fact is not in dispute—verifiable data have been compiled from the numerous experiments that have been conducted. As David Berlinski observed: "The cosmic hum is real enough, and so, too, is the fact that the universe is bathed in background radiation" (1998, p. 30). The ground data have been collected using the Caltech radio millimeter interferometer and the Owens Valley Array. Low-atmosphere instruments also have recorded CMB radiation using two balloon flights: MAXIMA (which, in 1998, flew at a height of approximately 24.5 miles for one night over Texas) and BOOMERANG (which, in 1998, flew at a height of around 23.5 miles for ten days over Antarctica), as well as from the Cosmic Background Explorer (COBE) and the Microwave Anisotropy Probe (MAP) satellite missions by NASA (Peterson, 1990; Flam, 1992; Musser, 2000).

What **is** in dispute is the **explanation** for the phenomenon. The late Sir Arthur Eddington—in his book, *The Internal Constitution of the Stars* (1926)—already had provided an accurate explanation for this temperature found in space. In the book's last chapter ("Diffuse Matter in Space"), he discussed the temperature in space. In Eddington's estimation, this phenomenon was not due to some ancient explosion, but rather was simply the background radiation from all of the heat sources that occupy the Universe. He calculated the minimum temperature to which any particular body in space would cool, given the fact that such bodies constantly are immersed in the radiation of distant starlight. With no adjustable parameters, he obtained a value of 3.18 K (later refined to 2.8)—essentially the same as the observed "background" radiation that is known to exist today.

In 1933, German scientist Erhard Regener showed that the intensity of the radiation coming from the plane of the Milky Way was essentially the same as that coming from a plane normal to it. He obtained a value of 2.8 K, which he felt would be the temperature characteristic of intergalactic space (Regener, 1933). His prediction came more than thirty years before Penzias and Wilson's discovery of the cosmic microwave background. The radiation that Big Bang theorists predicted was supposed to be much hotter than what was actually discovered. Gamow started his prediction at 5 K, and just a few years before Penzias and Wilson's discovery, suggested that it should be 50 K (see Alpher and Herman, 1949; Gamow, 1961a). As Van Flandern noted:

> The amount of radiation emitted by distant galaxies falls with increasing wavelengths, as expected if the longer wavelengths are scattered by the intergalactic medium. For example, the brightness ratio of radio galaxies at infrared and radio wavelengths changes with distance in a way which implies absorption. Basically, this means that the longer wavelengths are more easily absorbed by material between the galaxies. But then the microwave radiation (between the two wavelengths) should be absorbed by that medium too, and has no chance to reach us from such great distances,

or to remain perfectly uniform while doing so. **It must instead result from the radiation of microwaves from the intergalactic medium.** This argument alone implies that the microwaves could not be coming directly to us from a distance beyond all the galaxies, and therefore that the Big Bang theory cannot be correct.

None of the predictions of the background temperature based on the Big Bang was close enough to qualify as successes, the worst being Gamow's upward-revised estimate of 50 K made in 1961, just two years before the actual discovery. Clearly, without a realistic quantitative prediction, **the Big Bang's hypothetical "fireball" becomes indistinguishable from the natural minimum temperature of all cold matter in space** (2002, 9:73-74, parenthetical item in orig., emp. added).

Matter, whether on Earth or in space, absorbs radiation, and the CMB electromagnetic radiation is very likely the result of that absorption. Matter is known to absorb and emit radiation (known as blackbody radiation) caused by a change in temperature. Space is not an "empty" place, as some once thought, but is filled with stars, planets, nebulae, comets, asteroids, interstellar particles of dust and gas, and galaxies, all of which both absorb and emit varying amounts of radiation (see Akridge, et al., 1981, 18[3]:161). Fred Hoyle, Geoffrey Burbidge, and J.V. Narlikar, in their book, *A Different Approach to Cosmology* (2000), and Eric Lerner, in his book, *The Big Bang Never Happened* (1991), support the possibility of simple absorption and re-emission of the cosmic radiation. [Hoyle, et al., also suggested: "It seems very reasonable to suppose that the microwave radiation might very well have arisen from hydrogen burning in stars" (2000, p. 313).] Hoyle and his colleagues added to this thought when they stated that the "radiation field is generated by discrete objects and becomes smooth through scattering and diffusion in space" (p. 306). This, then, portrays a practical reason for the overall isotropy [spread out evenly in all directions] of the CMB radiation through thermalization and the scattering effect, also known as the Sunyaev-Zeldovich Effect (Humphreys, 1992, p. iii).

Despite their strong words of affirmation declaring to the world that they now had "proof," Big Bang supporters have had to admit that their theories about the CMB radiation are not really as concrete as they would like us to believe. Evolutionist Karen Fox confessed: "This radiation in and of itself doesn't require the big bang theory per se be correct" (2002, p. 134). Hoyle, et al., were a little more blunt: "...[T]he existence of the microwave radiation does not necessarily have anything to do with a big bang" (2000, p. 313). In fact, while the Big Bang Theory predicts that cosmic background radiation should exist, it does not necessarily predict that it should exist in thermal equilibrium. As Berlinski went on to note: "Although Big Bang cosmology does predict that the universe should be bathed in a milky film of radiation, it makes no predictions about the uniformity of its temperature" (1998, p. 30).

There was one thing, however, that cosmologists did recognize regarding the "uniformity of temperature" found in the background radiation. Initially, it represented a serious problem for the Big Bang Theory. It was "too" uniform—as science writers pointed out in articles with titles such as "Too Smooth a Universe" (see Folger, 1991). The formation of stars, galaxies, etc., during the early years of the Universe's formation, **required** that variations be present in the earliest distribution of the matter so that the matter ultimately would coalesce into those stars, galaxies, etc. And, as everyone acknowledged, the existence of these variations should have had some effect on the background radiation (see Lipkin, 1991, p. 23).

And that was the problem. When NASA sent up its COBE satellite in 1989, it found, at that time, a 3 K (or, to be more precise, a 2.735 ± 0.06 K) temperature—measured to an accuracy of 1 part in 10,000 (Peterson, 1990). In order for the early Universe to actually have formed in the manner in which they thought it did, scientists recognized that there **must** have been variations, however slight, in the background radiation. Yet, the background radiation seemed more pristine with each new look at the skies. Until 1992, the evidence of any serious fluctuations in the background radiation had been conspicuously absent, leaving the Big Bang concept riddled with problems for which there were seemingly no solutions (see Folger, 1991).

Perhaps you have heard that old saying: "That was then; this is now." Big Bang supporters now are suggesting that there is clear-cut evidence that the "cosmic egg" did, in fact, possess the necessary variations that allowed matter to coalesce into stars, galaxies, etc. A second survey was performed using NASA's COBE satellite, and was carried out to an accuracy, not of 1 in 10,000, but to 1 in 100,000 (see Flam, 1992). Astrophysicist George Smoot, and a team of scientists from the University of California at Berkeley, documented what seemed to be minor variations in the background temperature of the known Universe, thereby establishing the "fact" that there were variations present in the matter formed in the early stages of the Big Bang—variations that are presumed to represent the early defects that could explain how the Universe got to be so "lumpy" (see Smoot and Davidson, 1993). Smoot remarked to the Associated Press at the time, "If you're religious, it's like looking at God." On the front cover of Smoot's 1993 book, *Wrinkles in Time,* British astrophysicist Stephen W. Hawking is quoted as saying that the findings represent "the scientific discovery of the century, if not all time." And on the back cover of the book, the reader will find in big, bold, blue letters, "**Behold the handwriting of God**," followed by the statement: "George Smoot and his dedicated team of Berkeley researchers had proven the unprovable—uncovering, inarguably and for all time, the secrets of the creation of the Universe." WOW! Talk about fanfare!

In discussing the anisotropy of the radiation field, however, three things need to be considered. First, the temperature being measured is only a couple of degrees above absolute zero, —the point at which all motion ceases. Yet this radiation is alleged to have had its origin from an initial temperature of 10^{32} Celsius (Fox, p. 175). Second, most people likely are unaware of the infinitesimal nature of the variations being reported. In fact, these "variations" differ by barely **thirty-millionths** of a Kelvin! Some scientists doubt that these are big enough to account for the large-scale structure of the Universe (see Flam, 1992, 256:612). In an article titled "Boomerang Data Suggest a Purely Baryonic Universe" that he authored for *Astrophysics Journal,* astronomer Stacy McGaugh of the University of Maryland wrote:

> [C]osmic microwave background is very smooth. Structure cannot grow gravitationally to the rich extent seen today unless there is a non-baryonic component that can already be significantly clumped at the time of recombination without leaving **indiscriminately large fingerprints on the microwave background** (2000, 541:L33, emp. added).

But, as one scientist acknowledged, "the large fingerprints are just not observed" (Hartnett, 2001, 15[1]:10). Third, while the variations that have been measured have been documented in 1 part in **100,000**, cosmologists have stated that variations greater than 1 part in **10,000** are necessary for galaxies and clusters to form in the cosmological time that is allegedly available for gravity to carry out its work (see Rowan-Robinson, 1991).

Halton Arp likewise is skeptical of the significance of the new COBE results showing that the Universe displays a very slight anisotropy in the background radiation, which then is supposed to account for the rather clumpy distribution of matter in galaxies, superclusters, strings, etc. In his 1999 book, *Seeing Red: Redshifts, Cosmology and Academic Science,* Dr. Arp noted that in spite of these extremely slight irregularities of 1 part in 100,000, the background radiation is **still too smooth** to account for the clumpiness of the Universe (p. 237). The British journal, *Nature,* commented with subdued understatement: "The simple conclusion, that the data so far authenticated are consistent with the doctrine of the Big Bang, has been amplified in newspapers and broadcasts into proof that 'we now know' how the Universe began. This is cause for some alarm" (see "Big Bang Brouhaha," 1992, 356:731). There is indeed "cause for alarm." Allow me to explain.

With the assistance of a weather balloon, a telescope known as BOOMERANG spent ten days in December 1998 taking pictures of the Universe while flying over Antarctica. A few months earlier, a similar telescope called MAXIMA had flown high above Texas for a single night (see "MAXIMA, a Balloon-borne...," 2000). Both telescopes were designed to perform the exact same task, which was to observe the cosmic microwave radiation.

The telescopes were constructed to make precise maps of the "background radiation glow" on scales finer than one degree, which, according to researchers, would correspond to the size of the observable Universe at the time the radiation is thought to have been released. The design behind these experiments centered on the alleged random fluctuations (referred to as "hot" and "cold" spots) generated by cosmic in-

Figure 4 – Image at top left allegedly represents a "baby picture" of the Universe taken by the COBE satellite, first launched November 18, 1989. [Oval shape is a projection to display the entire sky, similar to the way the globe of the Earth can be projected as an oval.] Colors in the original images indicated "warmer" (red/yellow) and "cooler" (blue) spots. The image at the top right (taken by NASA's Wilkinson Microwave Anisotropy Probe [WMAP], launched June 30, 2001) brings the COBE picture into sharp focus, similar to refocusing a camera lense after taking an infant's snapshot, as in examples above. The high-resolution WMAP image supposedly depicts the microwave light from 380,000 years after the Big Bang, which is said to have occurred 13.7 billion years ago. This would be the equivalent of taking a picture of an 80-year-old man or woman on the day of his or her birth. CMB images courtesy of NASA.

flation in the first split second, which would have caused some regions of the Universe to be denser than others. As Ron Cowen summarized the matter in the September 28, 2002 issue of *Science News*: "The hot and cold spots represent the slightly uneven distribution of photons and matter in the early universe, which scientists view as the seeds of galaxy formation" (162: 195).

Supposedly, the telescopes could capture this difference in densities, which is said to have been caused by the ensuing battle between pressure and inertia that caused the plasma to oscillate between compression (an **increase** in density and pressure) and rarefaction (a **decrease** in density and pressure). As the Universe aged, so the theory goes, **oscillations** between compression and rarefaction developed on ever-larger scales. The fine detail in background radiation provided by these telescopes was supposed to provide a "snapshot" of the sound waves during those oscillations. Areas of compression would be somewhat hotter, thus brighter; areas of rarefaction would be cooler, thus darker. So, scientists spent many hours analyzing bright and dark areas captured by the telescopes.

At first, it appeared that the data fit quite nicely into researchers' theories. Cosmologist Michael S. Turner of the University of Chicago told a press conference in April 1999: "The Boomerang results fit the new cosmology like a glove" (as quoted in Musser, 283[1]:14). Additionally, a team of researchers, led by Paolo de Bernardis of the University of Rome, and Andrew E. Lange of the California Institute of Technology, declared in the April 27, 2000 issue of *Nature* that each of the BOOMERANG findings was "consistent with that expected for cold dark matter models in a flat Universe, as favoured by standard inflationary models" (de Bernardis, et al., 404:955). The MAXIMA team concluded similarly.

Once again, however, that was then, this is now. As it turns out, the images these two telescopes projected have challenged the very core of the Inflationary Big Bang Model itself. Three months after the *Nature* article appeared, George Musser penned an article ("Boomerang Effect") for the July 2000 issue of *Scientific American*, in which he wrote:

[W]hen measurements by the BOOMERANG and MAXIMA telescopes came in...scientists were elated.

...And then the dust settled, revealing that two pillars of big bang theory [the current status of the microwave background radiation and the necessity of a flat Universe—BT] were squarely in conflict.... That roar in the heavens may have been laughter at our cosmic confusion (283[1]:14,15).

Why is the Universe laughing at evolutionary cosmologists? What is this "confusion" all about? As Musser went on to explain, the BOOMERANG and MAXIMA telescopes

...made the most precise maps yet of the glow on scales finer than about one degree, which corresponds to the size of the observable universe at the time the radiation is thought to have been released (about 300,000 years after the bang). On this scale and smaller, gravity and other forces would have had enough time to sculpt matter.

For those first 300,000 years, the photons of the background radiation were bound up in a broiling plasma. Because of random fluctuations generated by cosmic inflation in the first split second, some regions happened to be denser. Their gravity sucked in material, whereupon the pressure imparted by the photons pushed that material apart again. The ensuing battle between pressure and inertia caused the plasma to oscillate between compression and rarefaction—vibrations characteristic of sound waves. As the universe aged, coherent oscillations developed on ever larger scales, filling the heavens with a deepening roar. But when the plasma cooled and condensed into hydrogen gas, the photons went their separate ways, and the universe abruptly went silent. **The fine detail in the background radiation is a snapshot of the sound waves at this instant** (283[1]:14, parenthetical items in orig., emp. added).

The data collected from BOOMERANG and MAXIMA were expected to show a profusion of different-sized spots—large spots would represent oscillations that had begun fairly recently, spots half that size would represent oscillations that had gone on for longer, spots a third that size would represent oscillations that had gone on longer still, and so on. Musser continued:

On either a Fourier analysis or a histogram of spot sizes, **this distribution would show up as a series of peaks**, each of which corresponds to the spots of a given size. The height of the peaks represents the maximum amount of compression (odd-numbered peaks) or of rarefaction (even-numbered peaks) in initially dense regions. Lo and behold, both telescopes saw the first peak [representing compression—BT]—which not only confirms that sounds reverberated through the early universe, as the big bang theory predicts, but also shows that the sounds were generated from preexisting fluctuations, as only inflation can produce (283[1]:14).

The data from both BOOMERANG and MAXIMA did indeed seem to be thrilling. Then, reality set in. The first significant problem with the information from the telescopes was that the data revealed only the "merest hint of a bulge where the second peak should be" (Musser, 283[1]:15). This was **really** bad news for inflationary theory, because it meant that the so-called "primordial plasma" contained numerous subatomic particles that weighed down the rarefaction of the sound waves and thereby suppressed the even-numbered peaks. Musser commented on the implication of this when he wrote:

According to Max Tegmark of the University of Pennsylvania and Matias Zaldarriaga of the Institute for Advanced Study in Princeton, N.J., **the Boomerang results imply that subatomic particles account for 50 percent more mass than standard big bang theory predicts—a difference 23 times larger than the error bars of the theory** (283[1]:15, emp. added).

Twenty-three times larger?! Whew! Where did those extra "subatomic particles" come from? No one knows. And inflationary theory cannot function with them present.

Just as the initial shock was beginning to wear off concerning the massive amounts of "extra subatomic particles" that the data revealed, more bad news began to pour in. Researchers needed (as required by inflationary cosmology) to find those "spots" (i.e., oscillations) moving outward and slightly upward at a very slight angle from an imaginary starting point on an imaginary flat plane (Euclidean geometry again—think "a sheet of paper"). The angle–according to the theory that is intended to predict a flat Universe–**could be no more than 0.8°**. The data from BOOMERANG, however, indicated an angle of **0.9°** (see Figure 5). If the Universe were flat, and if the rules of Euclidean trigonometry applied (both of which, the researchers agreed, would be the case), then the angle at which the "spots" propagated outward should have been no more than 0.8°.

Additional examination of the data revealed that this discrepancy in angles indicated that the Universe actually is **spherical**, not flat, because if anything starts out **completely** flat, then as it expands, it will not show curvature comparable to what the BOOMERANG telescope reported. As Musser wrote in *Scientific American*:

> ...[F]ollow-up studies soon showed that the lingering discrepancy, taken at face value, indicates that the universe is in fact spherical, with a density 10 percent greater than that required to make it flat. Such a gentle curvature seems awkward. Gravity quickly amplifies any deviations from exact flatness, so a slight sphericity today could only have arisen if the early universe was infinitesimally close to flat (283[1]:15, emp. added).

"Close to flat"–even "infinitesimally close to flat"–is not the same as "exact flatness." And therein lies the problem for inflationary theory. According to the BOOMERANG and MAXIMA data, then, there were too many subatomic particles present "in the beginning." And, to make matters worse, the Universe is spherical, not flat, as inflationary theory predicts.

Evolutionists (and those sympathetic with them) who have "put all their eggs into the inflationary theory basket" are understandably upset with the BOOMERANG and MAXIMA data and the obvious conclusions stemming from them, since, as Musser noted, this placed "two pillars of the big bang theory squarely in conflict." But the remaining alternatives are not much better. The only feasible alternative would seem to suggest that the trigonometric calculation used to account for "cosmic expansion"–couldn't! Such a scenario would occur only if: (1) the radiation did not travel as far as assumed (meaning it had been released **later** in cosmic history than expected);

(2) the famous Hubble constant were significantly larger (which would indicate that the Universe actually is **younger** than predicted); (3) the Universe contained more matter (which would hold back the expansion); or (4) the cosmological constant (discussed in detail later) were smaller (which would put the brakes on the current theory of cosmic acceleration).

And, unfortunately for Big Bang theorists, that **still** is not all the bad news. In its on-line "Science Update," *Nature* posted an article on Monday, March 31, 2003, titled "Sharp Images Blur Universal Picture." The author of that article, John Whitfield, remarked that

> physicists' notions of the Universe could be in trouble. New measurements from the Hubble Space Telescope hint that space is smooth, not grainy. Without graininess, our current theories predict that the Big Bang was infinitely hot and dense—tough to explain, to say the least (2003).

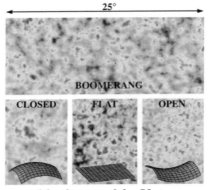

Figure 5 – The possible shapes of the Universe—closed, flat, or open—are based on how imaginary pairs of parallel lines might act. The bottom simulations represent the data that would result if each were correct, since BOOMERANG measures "hot" and "cold" spots (i.e., cosmic microwave background radiation) in the Universe. The top image depicts the **actual** BOOMERANG data.

If the Universe were "closed," the parallel lines eventually would converge upon each other (see bottom left). If the Universe were open, the parallel lines would diverge from each other (see bottom right). If the Universe were flat (like a sheet of paper), the parallel lines never would meet (see middle image).

"Tough to explain" happens to be another one of those "mild understatements." Richard Lieu of the University of Alabama at Huntsville (upon whose research Whitfield's report was partly based), admitted: "The theoreticians are very worried. There could be quite a lot of missing physics to be found" (as quoted in Whitfield). "Missing physics"? "Quite a lot" of "missing physics"? Robert Ragazzoni of the Astrophysical Observatory in Arcetri, Italy, agreed. "You don't see anything of the effect predicted" (as quoted in Whitfield). In short, things right now aren't looking very rosy for Big Bang inflationary theory. As nucleosynthesis expert David R. Tytler of the University of California at San Diego observed: "There are no known ways to reconcile these measurements and predictions" (as quoted in Musser, 283[1]:15).

Interestingly, not so long ago, adherents of the Big Bang held to a smooth Universe, and pointed with pride to the uniform background radiation. Then they found large-scale structures, and revised their predictions. Now, they have found infinitesimally small variations, and are hailing them as the greatest discovery of the twentieth century. But when a theory, claiming to be scientific, escapes falsification by continual modification with *ad hoc*, stopgap measures, caution is in order.

Let's face it: the Big Bang is a survivor. It **never** is falsified —only modified. David Lindley (1991) compared the efforts to revive existing cosmological theories with Ptolemy's work-around and fix-it solutions to an Earth-centered Solar System. Equations can be manipulated *ad infinitum* to make "messy" theories work, but Lindley warned, "skepticism is bound to arise."

And the skeptics are having a field day. In an article with a byline that reads like a *Who's Who* of Big Bang dissidents, Halton Arp and his allies have introduced a modified Steady State Theory. Not being able to resist taking a jab at their competitors, they wrote: "As a general scientific principle, it is undesirable to depend crucially on what is unobservable to explain what is observable, as happens frequently in Big Bang cosmology" (Arp, et al., 1990, 346:812). Elsewhere, Geoffrey Burbidge quipped: "To the zeroth order [at the simplest level—BT], the Big Bang is fine, but it doesn't account for the existence of us and stars, planets and galaxies" (as quoted in Peterson, 1991, 139:233). No, it certainly does not.

The Homogeneity of the Universe

The Big Bang model absolutely requires a uniform, homogeneous Universe. As I mentioned earlier, **isotropy** (matter being spread out evenly in all directions) and **homogeneity** (matter being spread out uniformly) are two foundational components of the standard Big Bang Theory. DePree and Axelrod addressed this fact when they wrote:

> Hubble made two very important discoveries in his studies of galaxy types and distributions. He found that the universe appeared to be both isotropic (the

Figure 6 – Representations of NASA's COBE and WMAP satellite probes, used to detect cosmic microwave background radiation. Images courtesy of NASA.

same in all directions), and homogeneous (one volume of space is much like any other volume of space). Together, the homogeneity and isotropy of the universe make up what we call the *cosmological principle:* **a cornerstone assumption in modern cosmology.** If we could not make this assumption (based on observation), then our cosmology might only apply to a very local part of the universe. But the cosmological principle allows us to extrapolate our conclusions drawn from our local viewpoint to the whole universe (2001, p. 363, parenthetical items and italics in orig., emp. added).

Berlinski summarized the critical need for homogeneity and isotropy in this manner:

In describing matter on a cosmic scale, cosmologists strip the stars and planets, the great galaxies and the bright bursting supernovae, of their uniqueness as places and things and replace them with an imaginary distribution: the matter of the universe is depicted as a great but uniform and homogeneous cloud covering the cosmos equitably in all its secret places. Cosmologists make this assumption because they must. There is no way to deal with the universe object by object; the equations would be inscrutable, impossible to solve.

Having simplified the contents of the universe, the cosmologist must take care as well, and for the same reason, to strip from the matter that remains any suggestion of particularity or preference in place. The universe, he must assume, is isotropic. It has no center whatsoever, no place toward which things tend, and no special direction or axis of coordination. The thing looks much the same wherever it is observed.

The twin assumptions that the universe is homogeneous and isotropic are not ancillary but indispensable to the hypothesis of an expanding universe; without them, no conclusion can mathematically be forthcoming (1998, pp. 34-35, emp. added).

But how, exactly, could the Big Bang account for the homogeneity that is supposed to exist within the Universe? That question, in fact, was one of six major problems with the standard Big Bang model that Andrei Linde discussed at length in his widely heralded November 1994 *Scientific American* article. Number five in that list was the following.

> Fifth, there is the question about the distribution of matter in the universe. On the very large scale, matter has spread out with remarkable uniformity. Across more than 10 billion light-years, its distribution departs from perfect homogeneity by less than one part in 10,000. For a long time, nobody had any idea why the universe was so homogeneous. But those who do not have ideas sometimes have principles. One of the cornerstones of the standard cosmology was the "cosmological principle," which asserts that the universe must be homogeneous. **This assumption, however, does not help much, because the universe incorporates important deviations from homogeneity, namely, stars, galaxies, and other agglomerations of matter. Hence, we must explain why the universe is so uniform on large scales and at the same time suggest some mechanism that produces galaxies** (1994, 271:49, emp. added).

The fact is, as Dr. Linde so eloquently pointed out, the Universe is "lumpy." **Really** lumpy! In a survey that covered one hundred-thousandth of the visible Universe, Margaret Geller and John Huchra (1989) identified a huge sheet-like structure that came to be called the "Great Wall." It contains thousands of galaxies, and extends at least 550 million light-years across the sky. Another survey, covering one two-thousandth of visible space, showed that the Universe **does** appear uniform— **but only on scales larger than 150 million light-years** (Cowen, 1990b).

As it turns out, there are at least two serious problems with any suggestion that the Universe exhibits homogeneity. First, homogeneity can be defended only if one considers the matter present in the Universe at distances greater than 150 million light-years. When it comes to getting "up close and personal," so to speak, the concept of homogeneity collapses completely —as Dr. Linde himself noted.

Second, a serious problem arises even when considering the matter of the Universe at distances greater than the 150-million-light-year cut-off point. A report by Saunders, et al. (1991), based on data from the Infrared Astronomical Satellite (IRAS), documented beyond doubt that there is more structure on large scales than is predicted by, or possible with, the standard cold dark matter theory of galaxy formation—which led the entire group of ten authors who performed the research and wrote the report to disavow completely the standard Big Bang theory. What shocked the scientific community was that the group included researchers who once were ardent supporters of the theory. The standard Big Bang Theory cannot account for the non-homogeneity of the Universe, which was Berlinski's point when he concluded: "However useful the assumption of homogeneity may be mathematically, it is false in the straightforward sense that the distribution of matter in the universe is not homogeneous at all" (p. 35, emp. added). Or, as Dr. Linde (quoted above) remarked with elegant understatement: "The universe incorporates important deviations from homogeneity." Yes, it does.

Dark Matter and Our "Precariously Balanced" Universe

In any Big Bang scenario—according to evolutionists' assumptions about the initial conditions—the Universe can contain no more than 10% protons, neutrons, and other ordinary matter found in stars, planets, galaxies, etc. What makes up the rest of the matter—90+% of the Universe—is still a mystery. As one physicist put it: "Astronomers therefore have no idea of the composition of the bulk of the entire universe. So much for a fundamental understanding of the physical universe" (De Young, 2000, 36:177).

Cosmologists do not know what the "mysterious stuff" is that composes "the bulk of the entire Universe." Nor have they found any credible, direct evidence of its existence. They refer to it as "cold dark matter" [CDM] (and/or "dark energy"—discussed later). As Stacy McGaugh wrote in *Astrophysics Journal:*

"As yet, we have no direct indication that CDM exists" (2000, 541:L33). A year later, John Hartnett wrote in agreement: "The dynamic behaviour of galaxies and galactic clusters begs for dark matter, as will be explained later, but to date, none has been found" (2001, 15[1]:9).

The mysterious and elusive "cold dark matter" is "cold" because it cannot interact with other matter (except gravitationally), and "dark" because it emits no detectable radiation, and therefore cannot be seen. In the March 2003 issue of *Scientific American,* David Cline authored an article titled "The Search for Dark Matter," in which he noted: "Being dark, it was never able to lose energy by emitting radiation, so it never could agglomerate into subgalactic clumps such as stars or planets" (288[3]:52). [In the scientific literature, cold dark matter also is referred to as "missing mass," "hidden matter," and "shadow matter."] Carl Sagan once described it as "dark, quintessential, deeply mysterious stuff wholly unknown on earth" (1994, p. 399). In his *Scientific American* article, Cline commented on this "unknown material" that makes up most of the Universe:

> **The terms we use** to describe its components, "dark matter" and "dark energy," **serve mainly as an expression of our ignorance**.... Essentially, all we know is that dark matter clumps together, providing a gravitational anchor for galaxies and larger structures such as galaxy clusters.... To detect dark matter, scientists need to know how it interacts with normal matter. Astronomers assume that it interacts only by means of gravitation, the weakest of all the known forces of nature. If that really is the case, physicists have no hope of ever detecting it (288[3]:52,54, emp. added).

Cline also remarked that even though, after seventy years of looking for it, we have no proof of the existence of dark matter, nevertheless, "nearly everyone accepts that it is real" (288 [3]:52). Why is this the case? The fact is, evolutionists **must** have this matter to support their theories. As DeYoung put it: "Popular versions of the big bang model **require immense amounts of dark matter** existing throughout space" (36:177, emp. added). Yes, they do, for two reasons. First, dark matter is necessary in order to allow for expansion and galaxy formation. If this "extra" matter did not exist, the ordinary matter of the Universe would have scattered into the empty reaches of space without ever coming together to form galaxies. Second, dark matter is mandatory for the success of the inflationary model of the origin of the Universe, and to ensure that the structure of the Universe is "flat," thereby guaranteeing that it will continue without end (concepts discussed below).

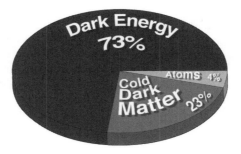

Figure 7 – Chart depicting the percentages of dark energy, dark matter, and actual matter (i.e., atoms) that must be present in order to explain the composition of the Universe via the Big Bang model

According to evolutionary cosmologists, the baffling yet profuse substance known as dark matter is present throughout the Universe, and, in fact, is the "invisible glue that holds it all together" (Lerner, 1991, p. 13; cf. DeYoung, 2000, 36:177). What is dark matter? DeYoung noted:

> This is an unanswered question since dark matter has never been directly observed, and may not even exist.... In reality, however, the dark matter mystery remains completely unsolved after seven decades of intense study (36:180,181).

Matter supposedly comes in a variety of types and forms: baryonic and non-baryonic, as well as cold and hot. Baryonic matter represents all the conventional matter (what Cline referred to as "normal matter") comprised of protons and neutrons. Non-baryonic dark matter is any matter not of a conventional nature–i.e., not composed of protons and neutrons. The "cold" and "hot" designations apply to this latter form only, and have to do with its motion [slow (cold) vs. fast (hot)] in relation to gravity. According to their own studies, evolutionists have concluded that the Universe is composed of a mere 4% baryonic matter, which **leaves 96% of the Universe as "dark" matter and/or "dark" energy.** In an article titled "Cosmology Gets Real" that appeared in the March 13, 2003 issue of *Nature*, staff writer Geoff Brumfiel wrote:

> With the addition of the latest data on the CMB [cosmic microwave background radiation–BT], courtesy of NASA's Wilkinson Microwave Anisotropy Probe, our picture of the universe is now clearer than ever. CMB studies have confirmed that the Universe is indeed flat. **The Wilkinson probe has now set ratios for the composition of the cosmos: 23% dark matter and 73% dark energy, leaving only 4% for the galaxies, stars and people** (422:109, emp. added).

Or, to echo the sentiments of cosmologist Michael Turner of the University of Chicago: "Ninety-six percent of the Universe is stuff that we've never seen" (as quoted in Brumfiel, 422:109) [see Figure 7].

Of the unseen Universe, dark matter is believed to constitute one third (33%) of its total mass (Milgrom, 2002, 287[2]: 44). And, "the galaxy motions suggest that the dark matter mass totals at least ten times that of all the visible galaxies" (DeYoung, 36:178). However, perhaps it would be wise to heed the evolutionists' own warning:

> Many suggestions have been made concerning the nature of the missing dark matter. Before embarking on flights of fancy, the reader should bear in mind that the astronomical evidence for a universe dominated by exotic forms of matter is slim, and the laboratory evidence for the various proposed candidates is equally slim. **Effective inflation, unless finely tuned, mandates the missing matter, yet we do not know what form it takes and so far have no evidence that it actually exists** (Harrison, 2000, p. 468, emp. added).

In his article in *Nature* on the character of the Cosmos, Brumfiel concluded: "...[T]he holes in our knowledge are still considerable. Researchers are confident that dark energy and dark matter are out there, but they don't know what kind of entities they are or how to find them" (422:109).

LONGHORN CONSTRUCTION SERVICES, INC.
General Contractors

9208 Lona Lane N.E.
Albuquerque, NM 87111

JAMES W. (BILL) LEWIS
President

5) 858-1360
(Fax) 858-1437
Email lcsinc@thuntek.net

But those "minor inconveniences" have not stopped those same researchers—in a last-ditch effort to establish the validity of their theories—from assigning actual percentages to the amount of dark matter that is supposed to exist, nor from giving specific names to its supposed forms. Some of these non-baryonic members allegedly include such eerie things as axions (named, believe it or not, after a laundry detergent!), WIMPs (weakly interacting massive particles), CHAMPs (Charged Massive Particles), and MACHOs (MAssive Compact Halo Objects) [Glashow, 1989; Palca, 1991; Silk, 1991]. Karen Fox admitted:

> The fact is that the dark matter problem is reaching something of a crisis, although few astronomers have been willing to admit this yet. Forget not finding any ideal dark matter candidates. The problem isn't that no one can find the missing matter (although they can't) but that even if theorists stomp their feet and shake their heads, observations haven't even shown that the universe is at the critical density (2002, pp. 122-123, parenthetical item in orig.).

But if "observations haven't even shown that the universe is at the critical density," then that plays havoc with the idea of inflation producing a Big-Bang-type of Universe that is flat, and that will expand indefinitely. As Fox casually remarked: "The dark matter problem affects the basics of the big bang model" (p. 124). It certainly does! John Gribbin confirmed such a position when he wrote that dark matter, "in a nutshell, is one of the biggest problems in cosmology today" (1981, pp. 315-316). Note the dates on these seemingly parallel statements. Interesting, is it not, that more than twenty years separate them, yet dark matter still "is one of the biggest problems in cosmology today"? [The reader may want to investigate the views of physicist Mordehai Milgrom of the Weizmann Institute of Science in Rehovot, Israel (see Milgrom, 2002). Dr. Milgrom has suggested that instead of opting for dark matter, cosmologists need to "re-tool the laws of physics," which he proposes to do via his Modified Newtonian Dynamics (MOND). Like American astronomer Halton Arp, Dr. Milgrom is viewed as somewhat of a heretic. In fact, "Dark-Matter Heretic" was the title of an article on the *American Scientist* Web site's "Science Observer" for January-February 2003 (see "Dark-Matter...").]

The fact is, the existence of dark matter is not merely a **theoretical prediction**, but rather a **necessary invention**—one that is required to fill the gaping holes in Big Bang cosmology and its cousin, inflationary theory (more about this shortly). Incredibly, **the hypothetical construct invented to *investigate* the theory has become the main support *for* the theory.** [As Berlinski put it: "The wish is father to the act" (1998, p. 31).] The importance of dark matter to evolutionary cosmology cannot be overstated. As Fox admitted: "Dropping dark matter out of their models would make it impossible for theorists to understand how a universe could get from the big bang to what it looks like today" (p. 124). Yes, it most definitely would, as Harnett went on to explain:

> These two issues [the existence of dark matter, and the microwave background radiation—BT] are fundamentally important to the evolutionary cosmologist. The missing dark matter in galaxies, galaxy clusters, and the whole universe, and the smoothness of the CMB radiation, create unassailable problems in the formation of stars and galaxies in the "early universe." ...The important questions left unanswered, of course, concern how stars and galaxies could have originated (2001, 15[1]:10).

On another front, an immense amount of time and energy has been expended in an attempt to determine the ultimate fate of the Universe. Will it collapse back on itself in a "Big Crunch," or will it simply continue expanding? Scientists have denoted the difference in these two—eventual contraction versus eternal expansion—as the Universe's "critical density." Simply put, if the mass density of the Universe itself is larger than the critical density, then gravity will prevail and the Universe allegedly will experience a Big Crunch. If the mass density of the Universe is lower than the critical density, then the Universe will expand forever, accelerating until it experiences a "Big Chill" (see Figure 8 on next page).

A third option is supposed to exist, however, when the mass density of the Universe is exactly equal to the critical density. According to scientists, this would allow the expansion of the

Universe to continue forever (even though the speed at which the Universe expands would decrease somewhat over time). To quote DeYoung:

> Dark matter is also involved in the popular inflationary big bang model which predicts that the curvature of the universe must be flat. This means that the density of matter is exactly balanced between a universe which eventually collapses (a closed, finite universe), and one which expands forever (an open, infinite universe). The required critical density for a flat universe is about 10^{-26} g/cm^3. This corresponds to approximately 10 hydrogen atoms per cubic meter of space. Observed density estimates, although crude, lead to a value 10-100 times smaller than the critical density. Therefore, a great amount of dark matter is needed to result in a flat, closed universe with zero curvature (2000, 36:180, parenthetical items in orig.).

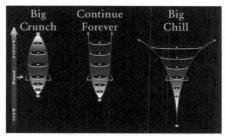

Figure 8 – Three models depicting the possible fate of the Universe from an evolutionary viewpoint. (1) In an expanding Universe, the combined gravity from the matter slows expansion. If the pull is strong enough, the expansion will stop and reverse itself, resulting in a "Big Crunch." (2) If the gravitational forces equal the expansion forces, then the Universe theoretically will continue forever (even though expansion slows down over time). (3) If gravitational forces are not strong enough, and are overcome by expansion forces, then the Universe supposedly will continue to expand, eventually ending in a "Big Chill."

In theory, scientists should be able to determine the fate of the Universe. In practical terms, however, there are major problems. One of the most important, as Dr. DeYoung has pointed out, is that there simply is not enough "ordinary" (observable) matter in the Universe to account for the observed gravitational forces that are holding galaxies together. Nor is there enough ordinary matter to ensure the "zero curvature" required by the inflationary concept (discussed in detail below) to guarantee the continued expansion of the Universe. Thus, in an attempt to salvage their naturalistic theories of the origin of the Universe, scientists simply invented dark matter. I use the word "invented" because dark matter is something that has been neither seen nor measured. As one scientist put it:

> So, cold dark matter is an unknown, unseen substance that is, nonetheless, essential to the process of self-creation.... Unfortunately, 90-99% of this matter is missing from the Universe. At this point, the Big Bang starts to bear striking similarities to the fable of the emperor's invisible new clothes (Major, 1991b, 11:23).

This is hardly an overstatement. An experimental report by French astronomers, Crézé, et al., in *Astronomy and Astrophysics* (1998), concluded that there is **no dark matter in the disk of the Milky Way Galaxy**. In commenting on the research, Alexander Hellemans wrote in *Science* shortly before the report by Crézé and his coworkers was published:

> By studying the movement of stars in the disk of our Milky Way galaxy, two teams of French astronomers have concluded that **what you see is what you get: The mass of the visible stars appears to account for all the material in the galactic disk**. These findings, derived from data gathered by the European astrometric satellite Hipparcos, imply that **the main body of our galaxy contains no "dark matter"**—invisible material that astronomers believe accounts for up to 90% of the mass of the universe (1997, 278:1230, emp. added).

Dr. Crézé and his colleagues analyzed the motion of stars perpendicular to the galactic disk in a sphere of radius 125 parsecs around the Sun. By analyzing the distribution of motion for 100 stars, the team was able to analyze the gravitational pull dragging them back toward the galactic disk. Why is this type of research important? *Nature* staff writer Brumfiel explained when he wrote in regard to dark matter:

The key to understanding it lies in its effects on stars and galaxies. According to general relativity, all mass distorts the space around it. When light from distant objects passes close to dark matter, it should be bent—a process called gravitational lensing.... Cosmologists also know a little about how dark matter interacts with other matter. The faster a particle moves, the more energy it transfers to any particles that it collides with. If, during the early Universe, dark matter was moving at close to the speed of light, it would have left its mark on the process by which matter clumped together to form stars and galaxies. **But astronomers can watch star and galaxy formation occurring in very distant parts of the Universe, and so far they have not seen any evidence of the influence of fast-moving dark matter** (2003, 422:109-110, emp. added).

The experimental research of Crézé, et al., agrees perfectly with Brumfiel's assessment—since the French team found **no evidence of fast-moving dark matter in the Milky Way Galaxy**.

Some might criticize the research of Crézé's team as being too small a sample in too small of a volume. Such criticism is muted, however, in a Ph.D. dissertation by Honc-Anh Pham of the Paris Observatory. She analyzed the motion of **10,000 stars** in the Milky Way disk (as opposed to Crézé's **100**). Pham's research produced a result similar to that of Crézé, et al. As Pham remarked: "These studies confirm that the **dark matter** [presumed to be] associated with the galactic disc **in fact doesn't exist**" (as quoted in Hellemans, 278:1230, emp. added).

One implication of this research could be that the Milky Way Galaxy is much younger than evolutionary astronomers believe. If our galaxy were representative of other galaxies, then it also would imply a much younger Universe as a whole. Have such astronomers abandoned the dark matter hypothesis and deduced a much younger Universe? Hardly! Instead, they merely have argued that the dark matter must be lurking in the **halo** of the Milky Way, rather than in the **disk**. The galactic halo is a large, spherical area that encircles the galaxy, and contains such things as dust, gas, and globular clusters. However, other scientists have debunked the idea that dark matter resides in the halo, and have concluded that the "dark chunks" previously reported in 1995 and 1996 (see Glanz, 1996) are very likely nothing but dim stars in the Magellanic Clouds (see Glanz, 1998, 281:332-333). Nathalie Palanque-Delabrouille of the Centre d'Études de Saclay in France concluded: "A halo interpretation of the other candidates becomes dubious" (as quoted in Glanz, 281:333). James Glanz, in reporting on this for *Science,* wrote: "One of astronomy's great mysteries, it seems, is still unsolved.... That's bad news for astronomers, who thought they finally had an answer to the puzzle of what could be holding galaxies together" (281:332,333).

The "other" bad news is–**that's not all the bad news!** Read on.

Dark Energy

As I noted previously, the concept of the Universe's expansion is critical to the Big Bang Theory and its cosmological cousin, Inflationary Theory. David Cline, in his March 2003 article on dark matter for *Scientific American,* noted: "Dark energy, despite its confusingly similar name [to dark matter–BT], is a separate substance that entered the picture only in 1998. It is spread uniformly through space, exerts a negative pressure and causes the expansion of the universe to accelerate" (288[3]:52). Geoff Brumfiel, writing in the March 13, 2003 issue of *Nature* about scientists' efforts to figure out **why** the Universe is expanding, observed that certain scientists have made

> an extraordinary suggestion: that the expansion of the Universe is accelerating, **pushed outwards by some kind of phantom force for which there was no explanation. This phenomenon of dark energy seemed odd**. But according to the general theory of relativity, mass and energy are equivalent. And when cosmologists looked at the amount of energy they needed to create the mysterious force, they found that it accounted perfectly for the mass still missing from their picture (422:109, emp. added).

Thus was born the idea of "dark energy." In the June 25, 2001 issue of *Time*, staff writer Michael Lemonick authored an article titled "The End," in which he commented: "...[A]strophysicists can be pretty sure they have assembled the full parts list for the cosmos at last: 5% ordinary matter, 35% exotic dark matter and about 60% dark energy" (157[25]:55). Astrophysicist John Barrow (co-author with Frank Tipler of the 1986 classic, *The Anthropic Cosmological Principle*) has suggested that the force of this dark energy is alleged to be "fifty per cent more than that of all the ordinary matter in the Universe" (2000, p. 191). That "dark energy" is the "phantom force" of which Brumfiel spoke. Or as science writer Paul Preuss remarked, it is an "an unknown form of energy often called the cosmological constant" (see Preuss, 2000).

Ah, yes–the famed "cosmological constant." Albert Einstein was the first to introduce the concept of the so-called cosmological constant–which he designated by the Greek letter Lambda (Λ)–to represent this "phantom force" or "unknown form of energy." It is–to be quite blunt–nothing more than a "fudge factor" set in place to make modern cosmology possible.

But this is no ordinary fudge factor. It happens to be, as Barrow correctly noted, "the smallest number ever encountered in science." And, as he observed, the value of lambda

> is bizarre: roughly 10^{-120}–that is, 1 divided by 10 followed by 119 zeros! This is the smallest number ever encountered in science. Why is it not zero? How can the minimum level be tuned so precisely? If it were 10 followed by just 117 zeros, then the galaxies could not form. **Extraordinary fine-tuning is needed to explain such extreme numbers**.... Why is its final state so close to the zero line? How does it "know" where to end up when the scalar field starts rolling downhill in its landscape? **Nobody knows the answers to these questions**. They are the greatest unsolved problems in gravitation physics and astronomy.... The only consolation is that, **if these observations are correct, there is now a very special value of lambda to try to explain** (pp. 259,260-261, emp. added).

A "very special value" indeed! Why is it so vanishingly small? Efstathiou, et al., writing on "The Cosmological Constant and Cold Dark Matter" in *Nature,* lamented:

> The cosmological term is a potential correction to the gravitational interaction. **If present at all, the cosmological term is incredibly small**: Its cumulative effects would show up only at the very largest length scales. However, **there is no compelling understanding of** *why* **the term is small** (1990, 348:705-707, emp. and italics added).

Nature's Brumfiel admitted:

> Dark energy is a more vexing problem, but the solution could lie in the nature of empty space. According to quantum theory, particles and their antiparticle equivalents are continually being created and annihilated, even in a vacuum. Some researchers have speculated that this vacuum energy could be what is accelerating the Universe's expansion. **But theoretical predictions for vacuum energy are up to 120 orders of magnitude greater than the strength of dark energy seen today** (2003, 422:110, emp. added).

Did Brumfiel say **120 orders of magnitude greater than the strength of dark energy** *seen today*? That implies that we have "seen" dark energy "today." But we have not! Similar to dark matter, "dark energy" is another mysterious concept that has been fabricated because the "theory still isn't jibing perfectly with observation" (Fox, p. 143). "Isn't jibing perfectly" is yet another magnificent understatement, considering that just previously, Fox had this to say concerning the present situation:

> For one thing, when the math was done to find what the cosmological constant should be via theory, **it was 10^{120} (that's a 1 followed by 120 zeros) times bigger than what we actually witness**. A cosmological constant that large would mean that everything in the universe should be expanding so quickly that you would not be able to see beyond the end of your nose (p. 143, parenthetical item in orig., emp. added).

What did Fox say—a 1 followed by 120 zeros? In the normal realm of science, that sort of error would be nothing short of catastrophic. No, on second thought, it would not even be scientific. Nobel Laureate Steven Weinberg, in his book *The First Three Minutes*, commented on this horrendous figure and its potential acceptance: "If we were to take this calculation seriously, it would undoubtedly be the most impressive quantitative disagreement between theory and experiment in the history of science!" (1977, p. 186). Or, to quote cosmologist Michael Turner: "Those models raise more questions than they answer. We've flushed out the basic features of the Universe. What we need now is a good story" (as quoted in Brumfiel, 422:110). "A good story" is exactly the foundation on which evolutionary cosmology has been constructed. It appears that Mark Twain was correct when he wrote in *Life on the Mississippi*: "There is something fascinating about science. One gets such wholesale returns of conjecture out of such a trifling investment of fact" (1883, p. 156).

DID THE UNIVERSE CREATE ITSELF OUT OF NOTHING?

In the February 2001 issue of *Scientific American*, Philip and Phylis Morrison authored an article titled "The Big Bang: Wit or Wisdom?," in which they remarked: "We no longer see a big bang as a direct solution" (284[2]:95). It's no wonder. As Andrei Linde also wrote in *Scientific American* (seven years earlier) about the supporting evidences for the Big Bang: "We found many to be highly suspicious" (1994, 271[5]:48).

Dr. Linde's comments caught no one by surprise—and drew no ire from his colleagues. In fact, long before he committed to print in such a prestigious science journal the Big Bang's obituary, cosmologists had known (though they were not too thrilled at the thought of having to admit it publicly) that the Big Bang is, to use my earlier phrase, "scientifically brain dead."

But it was because of that very fact that so many evolutionists had been working so diligently to find some way to "tweak" the Big Bang model so as to possibly revive it. As Berlinski rightly remarked:

Notwithstanding the investment made by the scientific community and the general public in contemporary cosmology, a suspicion lingers that matters do not sum up as they should. Cosmologists write as if they are quite certain of the Big Bang, yet, within the last decade, they have found it necessary to augment the standard view by means of various new theories. These schemes are meant to solve problems that cosmologists were never at pains to acknowledge, so that today they are somewhat in the position of a physician reporting both that his patient has not been ill and that he has been successfully revived (1998, p. 30).

Scientists are desperately searching for an answer that will allow them to continue to defend at least some form of the Big Bang Model. Berlinski went on to note:

Almost all cosmologists have a favored scheme; when not advancing their own, they occupy themselves enumerating the deficiencies of the others.... **Having constructed an elaborate scientific orthodoxy, cosmologists have acquired a vested interest in its defense**.... Like Darwin's theory of evolution, Big Bang cosmology has undergone that curious social process in which a scientific theory has been promoted to a secular myth (pp. 31-32,33,38, emp. added).

Enter inflationary theory—and the idea of (gulp!) a self-created Universe. In the past, it would have been practically impossible to find **any** reputable scientist who would have been willing to advocate a self-created Universe. To hold such a view would have been professional suicide. George Davis, a prominent physicist of the past generation, explained why when he wrote: "No material thing can create itself." Further, as Dr. Davis took pains to explain, such a statement "cannot be logically attacked on the basis of any knowledge available to us" (1958, p. 71). The Universe is the **created**, not the **Creator**. And until fairly recently, it seemed there could be no disagreement about that fact.

But, once again, "that was then; this is now." Because the standard Big Bang model is in such dire straits, and because the evidence is so conclusive that the Universe had some kind of beginning, evolutionists now are actually suggesting that **something came from nothing**–that is, **the Universe literally created itself from nothing**! Anthony Kenny, a British evolutionist, suggested in his volume, *Five Ways of Thomas Aquinas*, that something arose from nothing (1980). Edward P. Tryon, professor of physics at the City University of New York, was one of the first to suggest such an outlandish hypothesis: "In 1973," he said, "**I proposed that our Universe had been created spontaneously from nothing**, as a result of established principles of physics. This proposal variously struck people as preposterous, enchanting, or both" (1984, 101:14-16, emp. added). [This is the same Edward P. Tryon who went on record as stating: "Our universe is simply one of those things which happen from time to time" (1973, 246:397).]

Three years **earlier**, as it turned out, physicist Alan Guth of MIT had published a paper titled "Inflationary Universe: A Possible Solution to the Horizon and Flatness Problems," in which he outlined the specifics of inflationary theory (see Guth, 1981). Three years **later**, the idea that the Universe had simply "popped into existence from nothing," took flight when, in the May 1984 issue of *Scientific American*, Guth teamed up with physicist Paul Steinhardt of Princeton to co-author an article titled "The Inflationary Universe," in which they suggested:

> From a historical point of view probably the most revolutionary aspect of the inflationary model is the notion that all the matter and energy in the observable universe may have emerged from almost nothing.... The inflationary model of the universe provides a possible mechanism by which the observed universe could have evolved from an infinitesimal region. **It is then tempting to go one step further and speculate that the entire universe evolved from literally nothing** (1984, 250:128, emp. added).

Therefore, even though principles of physics that "cannot be logically attacked on the basis of any knowledge available to us" precluded the creation of something out of nothing, suddenly, in an eleventh-hour effort to resurrect the comatose Big Bang Theory, it was suggested that indeed, the Universe simply had "created itself out of nothing." As physicist John Gribbin suggested (in an article he wrote for *New Scientist* titled "Cosmologists Move Beyond the Big Bang") two years after Guth and Steinhardt offered their proposal: "...new models are based on the concept that particles [of matter–BT] can be created out of nothing at all...under certain conditions" and that "...matter might suddenly appear in large quantities" (1986, 110[1511]: 30).

Naturally, such a proposal would seem–to use Dr. Tryon's word–"preposterous." Be that as it may, some in the evolutionary camp were ready and willing to defend it–practically from the day it was suggested. One such scientist was Victor Stenger, professor of physics at the University of Hawaii. A mere three years after Guth and Steinhardt had published their volley in *Scientific American,* Dr. Stenger authored an article titled "Was the Universe Created?," in which he said:

> ...the universe is probably the result of a random quantum fluctuation in a spaceless, timeless void.... So what had to happen to start the universe was the formation of an empty bubble of highly curved space-time. How did this bubble form? What *caused* it? Not everything requires a cause. It could have just happened spontaneously as one of the many linear combinations of universes that has the quantum numbers of the void.... Much is still in the speculative stage, and **I must admit that there are yet no empirical or observational tests that can be used to test the idea of an accidental origin** (1987, 7[3]:26-30, italics in orig., emp. added.).

Not surprisingly, such a concept has met with rather stiff opposition from certain quarters within the scientific establishment. For example, in the summer 1994 edition of the *Skeptical Inquirer*, Ralph Estling wrote a stinging rebuke of the idea that the Universe created itself out of nothing. In his curiously titled article, "The Scalp-Tinglin', Mind-Blowin', Eye-Poppin', Heart-Wrenchin', Stomach-Churnin', Foot-Stumpin', Great Big Doodley Science Show!!!," Estling wrote:

The problem emerges in science when scientists leave the realm of science and enter that of philosophy and metaphysics, too often grandiose names for mere personal opinion, untrammeled by empirical evidence or logical analysis, and wearing the mask of deep wisdom.

And so they conjure us an entire Cosmos, or myriads of cosmoses, suddenly, inexplicably, causelessly leaping into being out of—out of Nothing Whatsoever, for no reason at all, and there-after expanding faster than light into more Nothing Whatsoever. And so cosmologists have given us Creation *ex nihilo*.... And at the instant of this Creation, they inform us, almost parenthetically, the universe possessed the interesting attributes of Infinite Temperature, Infinite Density, and Infinitesimal Volume, a rather gripping state of affairs, as well as something of a sudden and dramatic change from Nothing Whatsoever. They then intone equations and other ritual mathematical formulae and look upon it and pronounce it good.

I do not think that what these cosmologists, these quantum theorists, these universe-makers, are doing is science. I can't help feeling that universes are notoriously disinclined to spring into being, ready-made, out of nothing, even if Edward Tryon (ah, a name at last!) has written that "our universe is simply one of those things which happen from time to time...." Perhaps, although we have the word of many famous scientists for it, our universe is **not** simply one of those things that happen from time to time (18[4]:430, parenthetical item in orig., emp. added).

Estling's statements set off a wave of controversy, as was evident from subsequent letters to the *Skeptical Inquirer.* In the January/February 1995 edition of that journal, numerous letters were published, discussing Estling's article. Estling's response to his critics was published as well, and included the following observations:

All things begin with speculation, science not excluded. But if no empirical evidence is eventually forthcoming, or can be forthcoming, all speculation is barren.... **There is no evidence, so far, that the entire universe, observable and unobservable, emerged**

from a state of absolute Nothingness. Quantum cosmologists insist both on this absolute Nothingness and on endowing it with various qualities and characteristics: this particular Nothingness possesses virtual quanta seething in a false vacuum. Quanta, virtual or actual, false or true, are not Nothing, they are definitely Something, although we may argue over what exactly. For one thing, quanta are entities having energy, a vacuum has energy and moreover, extension, i.e., it is something into which other things, such as universes, can be put, i.e., we cannot have our absolute Nothingness and eat it too. If we have quanta and a vacuum as given, we in fact have a pre-existent state of existence that either pre-existed timelessly or brought itself into existence from absolute Nothingness (no quanta, no vacuum, no pre-existing initial conditions) at some precise moment in time; it creates this time, along with the space, matter, and energy, which we call the universe.... I've had correspondence with Paul Davies [a British astronomer who has championed the idea that the Universe created itself from nothing–BT] on cosmological theory, in the course of which, I asked him what he meant by "Nothing." He wrote back that he had asked Alexander Vilenkin what he meant by it and that Vilenkin had replied, "By Nothing I mean Nothing," which seemed pretty straightforward at the time, but these quantum cosmologists go on from there to tell us what their particular breed of Nothing consists of. I pointed this out to Davies, who replied that these things are very complicated. I'm willing to admit the truth of that statement, but I think **it does not solve the problem** (1995, 19[1]:69-70, emp. added).

This is an interesting turn of events. Evolutionists like Tryon, Stenger, Guth, and Steinhardt insist that this marvelously intricate Universe is "simply one of those things which happen from time to time" as the result of a "random quantum fluctuation in a spaceless, timeless void" that caused matter to evolve from "literally nothing." Such a suggestion, of course, would seem to be a clear violation of the first law of thermodynamics, which states that neither matter nor energy may be created or destroyed in nature. Berlinski acknowledged this when he wrote:

Hot Big Bang cosmology appears to be in violation of the first law of thermodynamics. The global energy needed to run the universe has come from nowhere, and to nowhere it apparently goes as the universe loses energy by cooling itself.

This contravention of thermodynamics expresses, in physical form, a general philosophical anxiety. Having brought space and time into existence, along with everything else, the Big Bang itself remains outside any causal scheme (1998, p. 37).

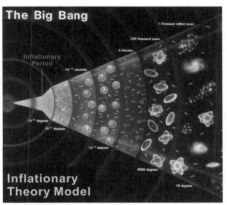

Figure 9 – Artist's depiction of the Big Bang Inflationary Model. Image courtesy of CERN.

But, as one might expect, supporters of inflation have come up with a response to that complaint, too. In discussing the Big Bang, Linde wrote in *Scientific American*:

In its standard form, the big bang theory maintains that the universe was born about 15 billion years ago from a cosmological singularity—a state in which the temperature and density are infinitely high. Of course, one cannot really speak in physical terms about these quantities as being infinite. **One usually assumes that the current laws of physics did not apply then** (1994, 271[5]:48).

Linde is not the only one willing to acknowledge what the essence of Big-Bang-type scenarios does to the basic laws of physics. Astronomer Joseph Silk wrote:

The universe began at time zero in a state of infinite density. Of course, the phrase "a state of infinite density" is completely unacceptable as a physical description of the universe.... **An infinitely dense universe [is] where the laws of physics, and even space and time, break down** (as quoted in Berlinski, 1998, p. 36).

But there are equally other serious problems as well. According to Guth, Steinhardt, Linde, and other evolutionary cosmologists, before the inflationary Big Bang, there was–well, **nothing**. Berlinski concluded: "But really the question of how the show started answers itself: before the Big Bang there was nothing" (p. 30). Or, as Terry Pratchett wrote: "The current state of knowledge can be summarized thus: In the beginning, there was nothing, which exploded" (1994, p. 7). Think about that for just a moment. Berlinski did, and then wrote:

The creation of the universe remains unexplained by any force, field, power, potency, influence, or instrumentality known to physics–or to man. **The whole vast imposing structure organizes itself from absolutely nothing. This is not simply difficult to grasp. It is incomprehensible.**

Physicists, no less than anyone else, are uneasy with the idea that the universe simply popped into existence, with space and time "suddenly switching themselves on." The image of a light switch comes from Paul Davies, who uses it to express a miracle without quite recognizing that it embodies a contradiction. **A universe that has suddenly switched itself on has accomplished something within time; and yet the Big Bang is supposed to have brought space and time into existence.**

Having entered a dark logical defile, physicists often find it difficult to withdraw. Thus, Alan Guth writes in pleased astonishment that the universe really did arise from "essentially...nothing at all": "as it happens, a false vacuum patch" "[10^{26}] centimeters in diameter" and "[10^{32}] solar masses." **It would appear, then, that "essentially nothing" has both spatial extension and mass. While these facts may strike Guth as inconspicuous, others may suspect that nothingness, like death, is not a matter that admits of degrees** (p. 37, emp. added).

In their more unguarded moments, evolutionary theorists admit as much. Writing in *Astronomy* magazine on "Planting Primordial Seeds," Rocky Kolb suggested: "In a very real sense, quantum fluctuations would be the origin of everything we see in the universe." Yet just one sentence prior to that, he admitted: "...[A] **region of seemingly empty space is not really empty, but is a seething froth** in which every sort of fundamental particle pops in and out of empty space before annihilating with its antiparticle and disappearing" (1998, 26 [2]:42,43, emp. added). Jonathan Sarfati commented:

> Some physicists assert that quantum mechanics violates this cause/effect principle and can produce something from nothing.... But this is a gross misapplication of quantum mechanics. **Quantum mechanics never produces something out of nothing**.... Theories that the Universe is a quantum fluctuation must presuppose that there was something to fluctuate—their "quantum vacuum" is a lot of matter-antimatter potential—not "nothing" (1998, 12[1]:21, emp. added).

Furthermore, as Kitty Ferguson has noted:

> Suppose it all began with a vacuum where space-time was empty and flat. The uncertainty principle doesn't allow an emptiness of complete zero.... In complete emptiness, the two measurements would read exactly zero simultaneously—zero value, zero rate of change—both very precise measurements. The uncertainty principle doesn't allow both measurements to be that definite at the same time, and therefore, as most physicists currently interpret the uncertainty principle, zero for both values simultaneously is out of the question. **Nothingness is forced to read—something**. If we can't have nothingness at the beginning of the universe, what do we have instead?...
>
> The "cosmological constant" is one of the values that seem to require fine-tuning at the beginning of the universe. You may recall from Chapter 4 that Einstein theorized about something called the "cosmological constant" which would offset the action of gravity in his theory, allowing the universe to remain static. Physicists now use the term to refer to the energy density of the vacuum. Common sense says there shouldn't be

any energy in a vacuum at all, but as we saw in Chapter 4, **the uncertainty principle doesn't allow empty space to be empty**....

Just as the uncertainty principle rules out the possibility of measuring simultaneously the precise momentum and the precise position of a particle, it also rules out the possibility of measuring simultaneously the value of a field and the rate at which that field is changing over time. The more precisely we try to measure one, the fuzzier the other measurement becomes. Zero is a very precise measurement, and measurement of two zeros simultaneously is therefore out of the question. Instead of empty space, there is a continuous fluctuation in the value of all fields, a wobbling a bit toward the positive and negative sides of zero so as not to *be* zero. **The upshot is that empty space instead of being empty must teem with energy** (1994, p. 171, italics in orig., emp. added).

Ultimately, the Guth/Steinhardt model for inflation was shown to be incorrect (see Guth and Weinberg, 1983), and a newer version was suggested. Working independently, Russian-American physicist Andrei Linde, and American physicists Andreas Albrecht and Paul Steinhardt, developed what came to be known as the "new inflationary model" (see Hawking, 1988, pp. 131-132; Linde, 1994, 271[5]:51). However, this model also was shown to be incorrect, and was discarded. Renowned British astrophysicist Stephen W. Hawking put the matter in proper perspective when he wrote:

The new inflationary model was a good attempt to explain why the universe is the way it is.... In my personal opinion, **the new inflationary model is now dead as a scientific theory**, although a lot of people do not seem to have heard of its demise and are still writing papers on it as if it were viable (1988, p. 132, emp. added).

Later, Linde suggested numerous modifications, and is credited with producing what became known as the "chaotic inflationary model" (see Hawking, pp. 132ff.). Dr. Hawking also performed additional work on this particular model. However, in an interview on June 8, 1994, dealing with inflationary models, Alan Guth conceded:

First of all, I will say that at the purely technical level, inflation itself does not explain how the universe arose from nothing.... Inflation itself takes a very small universe and produces from it a very big universe. But inflation by itself does not explain where that very small universe came from (as quoted in Heeren, 1995, p. 148).

After the chaotic inflationary model, came the eternal inflationary model, which was set forth by Linde in 1986. As Barrow summarized it in *The Book of Nothing:*

The spectacular effect of this is to make inflation self-reproducing. Every inflating region gives rise to other sub-regions which inflate and then in turn do the same. The process appears unstoppable—eternal. No reason has been found why it should ever end. Nor is it known if it needs to have a beginning. As with the process of chaotic inflation, every bout of inflation can produce a large region with very different properties. Some regions may inflate a lot, some only a little; some may have many large dimensions of space, some only three; some may contain four forces of Nature that we see, others may have fewer. The overall effect is to provide a physical mechanism by which to realize all, or at least almost all, possibilities somewhere within a single universe.

These speculative possibilities show some of the unending richness of the physicists' conception of the vacuum. It is the basis of our most successful theory of the Universe and why it has the properties that it does. Vacuums can change; vacuums can fluctuate; vacuums can have strange symmetries, strange geographies, strange histories. More and more of the remarkable features of the Universe we observe seem to be reflections of the properties of the vacuum (2000, pp. 256,271).

Michael J. Murray discussed the idea of the origin of the Universe via the Big Bang inflationary model.

According to the vacuum fluctuation models, our universe, along with these other universes, were generated by quantum fluctuations in a preexisting superspace. Imaginatively, one can think of this preexisting superspace as an infinitely extending ocean of soap, and each universe generated out of this superspace as a soap bubble which spontaneously forms on the ocean (1999, pp. 59-60).

Magnificent claims, to be sure—yet little more than wishful thinking. For example, cosmologists speak of a special particle —known as an "inflaton"—that is supposed to have provided the vacuum with its initial energy. Yet as scientists acknowledge, "...the particle that might have provided the vacuum energy density is still unidentified, even theoretically; it is sometimes called the inflaton because its sole purpose seems to be to have produced inflation" (see "The Inflationary Universe"). In an article on "Before the Big Bang" in the March 1999 issue of *Analog Science Fiction & Fact Magazine,* John Cramer wrote:

> The problem with all of this is that **the inflation scenario seems rather contrived and raises many unresolved questions**. Why is the universe created with the inflaton field displaced from equilibrium? Why is the displacement the same everywhere? What are the initial conditions that produce inflation? How can the inflationary phase be made to last long enough to produce our universe? Thus, **the inflation scenario which was invented to eliminate the contrived initial conditions of the Big Bang model apparently needs contrived initial conditions of its own** (1999, emp. added).

Cosmologist Michael Turner put it this way: "If inflation is the dynamite behind the Big Bang, we're still looking for the match" (as quoted in Overbye, 2001). Or, as journalist Dennis Overbye put it in an article titled "Before the Big Bang, There Was...What?" in the May 22, 2001 issue of *The New York Times:* "The only thing that all the experts agree on is that no idea works—yet" (2001). Barrow admitted somewhat sorrowfully: "So far, unfortunately, **the entire grand scheme of eternal inflation does not appear to be open to observational tests**" (2000, p. 256, emp. added). In his book, *The Accelerating Universe,* Mario Livio wrote in agreement:

> If eternal inflation really describes the evolution of the universe, then the beginning may be entirely inaccessible to observational tests. The point is that even

the original inflationary model, with a single inflation event, already had the property of erasing evidence from the preinflation epoch. **Eternal inflation appears to make any efforts to obtain information about the beginning, via observations in our own pocket universe, absolutely hopeless** (2000, pp. 180-181, emp. added).

Writing in the February 2001 issue of *Scientific American,* physicists Philip and Phylis Morrison admitted:

We simply do not know our cosmic origins; intriguing alternatives abound, but none yet compels. We do not know the details of inflation, nor what came before, nor the nature of the dark, unseen material, nor the nature of the repulsive forces that dilute gravity. The book of the cosmos is still open. Note carefully: **we no longer see a big bang as a direct solution. Inflation erases evidence of past space, time and matter.** The beginning—if any—is still unread (284[2]:95, emp. added).

But Dr. Barrow went even farther:

As the implications of the quantum picture of matter were explored more fully, a further radically new consequence appears that was to impinge upon the concept of the vacuum. Werner Heisenberg showed that there were complementary pairs of attributes of things which could not be measured simultaneously with arbitrary precision, even with perfect instruments. This restriction on measurement became known as the Uncertainty Principle. One pair of complementary attributes limited by the Uncertainty Principle is the combination of position and momentum. Thus we cannot know at once where something is *and* how it is moving with arbitrary precision....

The Uncertainty Principle and the quantum theory revolutionised our conception of the vacuum. **We can no longer sustain the simple idea that a vacuum is just an empty box.** If we could say that there were no particles in a box, that it was completely empty of all mass and energy, then we would have to violate the Uncertainty Principle because we would require perfect information about motion at every point and about the energy of the system at a given instant of time....

This discovery at the heart of the quantum description of matter means that the concept of a vacuum must be somewhat realigned. **It is no longer to be associated with the idea of the void and of nothingness or empty space. Rather, it is merely the emptiest possible state in the sense of the state that possesses the lowest possible energy; the state from which no further energy can be removed** (2000, pp. 204,205, italics in orig.; emp. added).

The simple fact is, to quote R.C. Sproul, "Every **effect** must have a **cause**. That is true by definition.... It is impossible for something to create itself. The concept of self-creation is a contradiction in terms, a nonsense statement.... [S]elf-creation is irrational" (1992, p. 37, emp. in orig.).

Furthermore, science is based on observation, reproducibility, and empirical data. But when pressed for the empirical data that document the claim that the Universe created itself from nothing, evolutionists are forced to admit, as Dr. Stenger did, that "...there are yet no empirical or observational tests that can be used to test the idea...." Estling summarized the problem quite well when he stated: "There is no evidence that the entire universe, observable and unobservable, emerged from a state of absolute Nothingness." Agreed.

WAS THE UNIVERSE CREATED?

The Universe is not eternal. Nor did it create itself. It therefore must have been created. And such a creation most definitely implies a Creator.

Is the Universe the result of creation by an eternal Creator? Either the Universe had a beginning, or it had no beginning. But all available evidence asserts that the Universe did have a beginning. If the Universe had a beginning, it either had a cause, or it did not have a cause. One thing we know: it is correct—both scientifically and philosophically—to acknowledge that the Universe had an adequate cause, because the Universe is an effect, and as such requires an adequate antecedent cause. Nothing causeless happens. Henry Morris was entirely correct when he suggested that the Law of Cause and Effect is "universally accepted and followed in every field of science" (1974, p. 19). The cause/effect principle states that wherever there is a material **effect**, there must be an adequate antecedent **cause**. Further indicated, however, is the fact that no effect can be qualitatively superior to, or quantitatively greater than, its cause.

Since it is apparent that the Universe is not eternal, and since it likewise is apparent that the Universe could not have created itself, the only remaining alternative is that the Universe **was created** by something (or Someone): (a) that existed before it, i.e., some eternal, uncaused First Cause; (b) superior to it—the created cannot be superior to the creator; and (c) of a different nature since the finite, dependent Universe of matter is unable to explain itself. As Hoyle and Wickramasinghe observed: "To be consistent logically, we have to say that the intelligence which assembled the enzymes did not itself contain them" (1981, p. 139).

In connection with this, another fact should be considered. If there ever had been a time when absolutely **nothing** existed, then there would be nothing now. It is a self-evident truth that nothing produces nothing. In view of this, **since something does exist, it must follow logically that something has existed forever!** Everything that exists can be classified as either **matter** or **mind**. There is no third alternative. The argument then, is this:

1. Everything that exists is either matter or mind.

2. Something exists now, so something eternal exists.

3. Therefore, either matter or mind is eternal.

A. Either matter or mind is eternal.

B. Matter is not eternal, per the evidence cited above.

C. Thus, it is mind that is eternal.

Or, to reason somewhat differently:

1. Everything that is, is either dependent (i.e., contingent) or independent (non-contingent).
2. If the Universe is not eternal, it is dependent (contingent).
3. The Universe is not eternal.
4. Therefore, the Universe is dependent (contingent).
A. If the Universe is dependent, it must have been caused by something that is independent.
B. But the Universe is dependent (contingent).
C. Therefore, the Universe was produced by some eternal, independent (non-contingent) force.

In the past, atheistic evolutionists suggested that the mind is nothing more than a function of the brain, which is matter; thus the mind and the brain are the same, and matter is all that exists. As the late evolutionist of Cornell University, Carl Sagan, said in the opening sentence of his television extravaganza (and book by the same name), *Cosmos*, "The Cosmos is all that is or ever was or ever will be" (1980, p. 4). However, that viewpoint no longer is credible scientifically, due in large part to the experiments of Australian physiologist Sir John Eccles. Dr. Eccles, who won the Nobel Prize for his discoveries relating to the neural synapses within the brain, documented that the mind is more than merely physical. He showed that the supplementary motor area of the brain may be fired by mere **intention** to do something, without the motor cortex (which controls muscle movements) operating. In effect, the mind is to the brain what a librarian is to a library. The former is not reducible to the latter. Eccles explained his methodology and conclusions in *The Self and Its Brain*, co-authored with the renowned philosopher of science, Sir Karl Popper (see Popper and Eccles, 1977), as well as in a number of other volumes that he authored.

Anyone familiar with neurophysiology or neurobiology knows the name of Sir John Eccles. But for those who might not be familiar with this amazing gentleman, I would like to introduce Dr. Eccles via the following quotation, which is from a chapter ("The Collapse of Modern Atheism") that philosopher Norman Geisler authored for the book, *The Intellectuals Speak Out About God* (which also contained a chapter by Eccles, from which I will quote shortly). Geisler wrote:

> The extreme form of materialism believes that mind (or soul) *is* matter. More modern forms believe mind is *reducible to* matter or *dependent on* it. **However, from a scientific perspective much has happened in our generation to lay bare the clay feet of materialism. Most noteworthy among this is the Nobel Prize winning work of Sir John Eccles. His work on the brain demonstrated that the mind or intention is more than physical. He has shown that the supplementary motor area of the brain is fired by mere *intention* to do something, without the motor cortex of the brain (which controls muscle movements) operating**. So, in effect, the mind is to the brain what an archivist is to a library. The former is not reducible to the latter (1984, pp. 140-141, parenthetical items and italics in orig., emp. added).

Eccles and Popper viewed the mind as a distinctly non-material entity. But neither did so for religious reasons, as both were committed Darwinians. Rather, they believed what they did about the human mind because of their own research. Eccles spent his entire adult life studying the brain-mind problem, and concluded that the two were entirely separate entities. In a fascinating book, *Nobel Conversations*, Norman Cousins, who moderated a series of conversations among four Nobel laureates, including Dr. Eccles, made the following statement: "Nor was Sir John Eccles claiming too much when **he insisted that the action of non-material mind on material brain has been not merely postulated but scientifically demonstrated**" (1985, p. 68, emp. added). Eccles himself, in his book, *The Understanding of the Brain*, wrote:

> When I postulated many years ago, following Sherrington [Sir Charles Sherrington, Nobel laureate and Eccles' mentor—BT], that there was a special area of

the brain in liaison with consciousness, I certainly did not imagine that any definitive experimental test could be applied in a few years. But now we have this distinction between the dominant hemisphere in liaison with the conscious self, and the minor hemisphere with no such liaison (1973, p. 214).

In an article—"Scientists in Search of the Soul"—that examined the groundbreaking work of Dr. Eccles (and other scientists like him who have been studying the mind/brain relationship), science writer John Gliedman wrote:

At age 79, Sir John Eccles is not going "gentle into the night." Still trim and vigorous, the great physiologist has declared war on the past 300 years of scientific speculation about man's nature.

Winner of the 1963 Nobel Prize in Physiology or Medicine for his pioneering research on the synapse—the point at which nerve cells communicate with the brain —Eccles strongly defends the ancient religious belief that human beings consist of a mysterious compound of physical and intangible spirit.

Each of us embodies a nonmaterial thinking and perceiving self that "entered" our physical brain sometime during embryological development or very early childhood, says the man who helped lay the cornerstones of modern neurophysiology. This "ghost in the machine" is responsible for everything that makes us distinctly human: conscious self-awareness, free will, personal identity, creativity and even emotions such as love, fear, and hate. Our nonmaterial self controls its "liaison brain" the way a driver steers a car or a programmer directs a computer. Man's ghostly spiritual presence, says Eccles, exerts just the whisper of a physical influence on the computerlike brain, enough to encourage some neurons to fire and others to remain silent. Boldly advancing what for most scientists is the greatest heresy of all, Eccles also asserts that our nonmaterial self survives the death of the physical brain (1982, 90[7]:77).

While discussing the same type of conclusions reached by Dr. Eccles, Geisler explored the concept of an eternal, all-knowing Mind.

Further, this infinite cause of all that is must be all-knowing. It must be knowing because knowing beings exist. I am a knowing being, and I know it. I cannot meaningfully deny that I can know without engaging in an act of knowledge.... But a cause can communicate to its effect only what it has to communicate. If the effect actually possesses some characteristic, then this characteristic is properly attributed to its cause. The cause cannot give what it does not have to give. If my mind or ability to know is received, then there must be Mind or Knower who gave it to me. The intellectual does not arise from the nonintellectual; something cannot arise from nothing. The cause of knowing, however, is infinite. Therefore it must know infinitely. It is also simple, eternal, and unchanging. Hence, whatever it knows—and it knows anything it is possible to know—it must know simply, eternally, and in an unchanging way (1976, p. 247).

From such evidence, Robert Jastrow concluded: "That there are what I or anyone would call supernatural forces at work is now, I think, a scientifically proven fact..." (1982, p. 18). Apparently Dr. Jastrow is not alone. As Gliedman put it:

Eccles is not the only world-famous scientist taking a controversial new look at the ancient mind-body conundrum. From Berkeley to Paris and from London to Princeton, prominent scientists from fields as diverse as neurophysiology and quantum physics are coming out of the closet and admitting they believe in the possibility, at least, of such unscientific entities as the immortal human spirit and divine creation (90 [7]:77).

In an article titled "Modern Biology and the Turn to Belief in God" that he wrote for the book, *The Intellectuals Speak Out About God,* Eccles concluded:

Science and religion are very much alike. Both are imaginative and creative aspects of the human mind. The appearance of a conflict is a result of ignorance. **We come to exist through a divine act.** That divine guidance is a theme throughout our life; at our death the brain goes, but that divine guidance and love continues. Each of us is a unique, conscious being, a divine creation. **It is the religious view. It is the only view consistent with all the evidence....**

Since materialist solutions fail to account for our experienced uniqueness, **we are constrained to attribute the uniqueness of the psyche to a supernatural creation**. To give the explanation in theological terms: Each soul is a Divine creation, which is "attached" to the growing fetus at some point between conception and birth. It is the certainty of the inner core of unique individuality that necessitates the "Divine creation." We submit that no other explanation is tenable (1984, pp. 43,50, emp. added).

Or, as Jastrow lamented:

For the scientist who has lived by his faith in the power of reason, the story ends like a bad dream. He has scaled the mountains of ignorance; he is about to conquer the highest peak; as he pulls himself over the final rock, he is greeted by a band of theologians who have been sitting there for centuries (1978, p. 116).

Our Fine-Tuned, Tailor-Made Universe

And it is not just people who are unique (in the sense of exhibiting evidence of design). The fact is, the Universe is "fine-tuned" in such a way that it is impossible to suggest logically that it simply "popped into existence out of nothing" and then went from the chaos associated with the inflationary Big Bang model (as if the Universe were a giant firecracker!) to the sublime order that it presently exhibits. Murphy and Ellis went on to note:

The symmetries and delicate balances we observe in the universe require an extraordinary coherence of conditions and cooperation of laws and effects, suggesting that in some sense they have been **purposely designed**. That is, **they give evidence of intention**, realized both in the setting of the laws of physics and in the choice of boundary conditions for the universe (p. 57, emp. added).

The idea that the Universe and its laws "have been purposely designed" has surfaced much more frequently in the past several years. For example, Sir Fred Hoyle wrote:

A common sense interpretation of the facts suggests that a superintellect has monkeyed with physics, as well as with chemistry and biology, and that there are no blind forces worth speaking about in nature. The numbers one calculates from the facts seem to me so overwhelming as to put this conclusion almost beyond question (1982, 20:16).

In his book, *Superforce: The Search for a Grand Unified Theory of Nature*, Australian astrophysicist Paul Davies made this amazing statement:

If nature is so "clever" as to exploit mechanisms that amaze us with their ingenuity, **is that not persuasive evidence for the existence of intelligent design behind the universe**? If the world's finest minds can unravel only with difficulty the deeper workings of nature, how could it be supposed that those workings are merely a mindless accident, a product of blind chance? (1984, pp. 235-236, emp. added).

Four years later, in his text, *The Cosmic Blueprint: New Discoveries in Nature's Creative Ability to Order the Universe*, Davies went even further when he wrote:

There is for me powerful evidence that there is something going on behind it all.... **It seems as though somebody has fine-tuned nature's numbers to make the Universe.... The impression of design is overwhelming** (1988, p. 203, emp. added).

Another four years later, in 1992, Davies authored *The Mind of God*, in which he remarked:

I cannot believe that our existence in this universe is a mere quirk of fate, an accident of history, an incidental blip in the great cosmic drama.... Through conscious beings the universe has generated self-awareness. This can be no trivial detail, no minor by-product of mindless, purposeless forces. **We are truly meant to be here** (1992, p. 232, emp. added).

That statement, "We are truly meant to be here," was the type of sentiment expressed by two scientists, John Barrow and Frank Tipler, in their 1986 book, *The Anthropic Cosmological Principle*, which discussed the possibility that the Universe seems to have been "tailor-made" for man. Eight years after that book was published, Dr. Tipler wrote *The Physics of Immortality*, in which he professed:

When I began my career as a cosmologist some twenty years ago, I was a convinced atheist. I never in my wildest dreams imagined that one day I would be writing a book purporting to show that the central claims of Judeo-Christian theology are in fact true, that these claims are straightforward deductions of the laws of physics as we now understand them. I have been forced into these conclusions by the inexorable logic of my own special branch of physics (1994, Preface).

In 1995, NASA astronomer John O'Keefe stated in an interview:

We are, by astronomical standards, a pampered, cosseted, cherished group of creatures.... If the Universe had not been made with the most exacting precision we could never have come into existence. It is my view that these circumstances indicate the universe was created for man to live in (as quoted in Heeren, 1995, p. 200).

Then, thirteen years after he published his 1985 book (*Evolution: A Theory in Crisis*), Michael Denton shocked everyone —especially his evolutionist colleagues—when he published his 1998 tome, *Nature's Destiny*, in which he admitted:

Whether one accepts or rejects the design hypothesis... there is no avoiding the conclusion that the world **looks** as if it has been tailored for life; it **appears to have been designed. All reality appears** to be a vast, coherent, teleological whole with life and mankind as its purpose and goal (p. 387, emp. in orig.).

In his discussion of the Big Bang inflationary model, Murray discussed the idea of the origin of the Universe and the complexity that would be required to pull off such an event.

...[I]n all current worked-out proposals for what this "universe generator" could be—such as the oscillating big bang and the vacuum fluctuation models explained above—the "generator" itself is governed by a complex set of physical laws that allow it to produce the universes. It stands to reason, therefore, that if these laws were slightly different the generator probably would not be able to produce any universes that could sustain life. After all, even my bread machine has to be made just right to work properly, and it only produces loaves of bread, not universes!

...[T]he universe generator must not only select the parameters of physics at random, but must actually randomly create or select the very laws of physics themselves. This makes this hypothesis seem even more far-fetched since it is difficult to see what possible physical mechanism could select or create such laws. The reason the "many-universes generator" must randomly select the laws of physics is that, just as the right values for the parameters of physics are needed for life to occur, the right set of laws is also needed. If, for instance, certain laws of physics were missing, life would be impossible. For example, without the law of inertia, which guarantees that particles do not shoot off at high speeds, life would probably not be possible. Another example is the law of gravity; if masses did not attract each other, there would be no planets or stars, and once again it seems that life would be impossible (1999, pp. 61-62).

Hoyle addressed the fine-tuning of the nuclear resonances responsible for the oxygen and carbon synthesis in stars when he observed:

I do not believe that any scientists who examined the evidence would fail to draw the inference that **the laws of nuclear physics have been deliberately designed** with regard to the consequences they produce inside stars. If this is so, then my apparently random quirks have become part of a deep-laid scheme. If not, then we are back again at a **monstrous sequence of accidents** (1959, emp. added).

When we (to use Hoyle's words) "examine the evidence," what do we find? Murray answered:

Almost **everything about the basic structure of the universe**—for example, the fundamental laws and parameters of physics and the initial distribution of matter and energy—**is balanced on a razor's edge** for life to occur.... Scientists call this extraordinary balancing of the parameters of physics and the initial conditions of the universe the "fine-tuning of the cosmos" (1999, p. 48, emp. added).

Indeed they do. And it is fine-tuning to a remarkable degree. Consider the following critically important parameters that must be fine-tuned (from an evolutionary perspective) in order for the Universe to exist, and for life to exist in the Universe.

1. Strong nuclear force constant:

if larger: no hydrogen would form; atomic nuclei for most life-essential elements would be unstable; thus, no life chemistry;

if smaller: no elements heavier than hydrogen would form: again, no life chemistry

2. Weak nuclear force constant:

if larger: too much hydrogen would convert to helium in big bang; hence, stars would convert too much matter into heavy elements making life chemistry impossible;

if smaller: too little helium would be produced from the big bang; hence, stars would convert too little matter into heavy elements making life chemistry impossible

3. Gravitational force constant:

if larger: stars would be too hot and would burn too rapidly and too unevenly for life chemistry;

if smaller: stars would be too cool to ignite nuclear fusion; thus, many of the elements needed for life chemistry would never form

4. Electromagnetic force constant:

if greater: chemical bonding would be disrupted; elements more massive than boron would be unstable to fission;

if lesser: chemical bonding would be insufficient for life chemistry

5. Ratio of electromagnetic force constant to gravitational force constant:

if larger: all stars would be at least 40% more massive than the Sun; hence, stellar burning would be too brief and too uneven for life support;

if smaller: all stars would be at least 20% less massive than the Sun, thus incapable of producing heavy elements

6. Ratio of electron to proton mass:

if larger: chemical bonding would be insufficient for life chemistry;

if smaller: same as above ratio of number of protons to number of electrons

7. Ratio of number of protons to number of electrons:

if larger: electromagnetism would dominate gravity, preventing galaxy, star, and planet formation;

if smaller: same as above

8. Expansion rate of the Universe:

if larger: no galaxies would form

if smaller: Universe would collapse, even before stars formed entropy level of the Universe

9. Entropy level of the Universe:

if larger: stars would not form within proto-galaxies;

if smaller: no proto-galaxies would form

10. Mass density of the Universe:

if larger: overabundance of deuterium from big bang would cause stars to burn rapidly, too rapidly for life to form;

if smaller: insufficient helium from big bang would result in a shortage of heavy elements

11. Velocity of light:

if faster: stars would be too luminous for life support;

if slower: stars would be insufficiently luminous for life support

12. Initial uniformity of radiation:

if more uniform: stars, star clusters, and galaxies would not have formed;

if less uniform: Universe by now would be mostly black holes and empty space

13. Average distance between galaxies:

if larger: star formation late enough in the history of the Universe would be hampered by lack of material

if smaller: gravitational tug-of-wars would destabilize the Sun's orbit

14. Density of galaxy cluster:

if denser: galaxy collisions and mergers would disrupt the sun's orbit

if less dense: star formation late enough in the history of the universe would be hampered by lack of material

15. Average distance between stars:

if larger: heavy element density would be too sparse for rocky planets to form

if smaller: planetary orbits would be too unstable for life

16. Fine structure constant (describing the fine-structure splitting of spectral lines):

if larger: all stars would be at least 30% less massive than the Sun

if larger than 0.06: matter would be unstable in large magnetic fields

if smaller: all stars would be at least 80% more massive than the Sun

17. Decay rate of protons:

if greater: life would be exterminated by the release of radiation

if smaller: Universe would contain insufficient matter for life

18. ^{12}C to ^{16}O nuclear energy level ratio:

if larger: Universe would contain insufficient oxygen for life

if smaller: Universe would contain insufficient carbon for life

19. Ground state energy level for ^{4}He:

if larger: Universe would contain insufficient carbon and oxygen for life

if smaller: same as above

20. Decay rate of ^{8}Be:

if slower: heavy element fusion would generate catastrophic explosions in all the stars

if faster: no element heavier than beryllium would form; thus, no life chemistry

21. Ratio of neutron mass to proton mass:

if higher: neutron decay would yield too few neutrons for the formation of many life-essential elements

if lower: neutron decay would produce so many neutrons as to collapse all stars into neutron stars or black holes

22. Initial excess of nucleons over anti-nucleons:

if greater: radiation would prohibit planet formation

if lesser: matter would be insufficient for galaxy or star formation

23. Polarity of the water molecule:

if greater: heat of fusion and vaporization would be too high for life

if smaller: heat of fusion and vaporization would be too low for life; liquid water would not work as a solvent for life chemistry; ice would not float, and a runaway freeze-up would result

24. Supernovae eruptions:

if too close, too frequent, or too late: radiation would exterminate life on the planet

if too distant, too infrequent, or too soon: heavy elements would be too sparse for rocky planets to form

25. White dwarf binaries:

if too few: insufficient fluorine would exist for life chemistry

if too many: planetary orbits would be too unstable for life

if formed too soon: insufficient fluorine production

if formed too late: fluorine would arrive too late for life chemistry

26. Ratio of exotic matter mass to ordinary matter mass:

if larger: universe would collapse before solar-type stars could form

if smaller: no galaxies would form

27. Number of effective dimensions in the early Universe:

if larger: quantum mechanics, gravity, and relativity could not coexist; thus, life would be impossible

if smaller: same result

28. Number of effective dimensions in the present Universe:

if smaller: electron, planet, and star orbits would become unstable

if larger: same result

29. Mass of the neutrino:

if smaller: galaxy clusters, galaxies, and stars would not form

if larger: galaxy clusters and galaxies would be too dense

30. Big bang ripples:

if smaller: galaxies would not form; Universe would expand too rapidly:

if larger: galaxies/galaxy clusters would be too dense for life; black holes would dominate; Universe would collapse before life-site could form

31. Size of the relativistic dilation factor:

if smaller: certain life-essential chemical reactions will not function properly

if larger: same result

32. Uncertainty magnitude in the Heisenberg uncertainty principle:

if smaller: oxygen transport to body cells would be too small and certain life-essential elements would be unstable

if larger: oxygen transport to body cells would be too great and certain life-essential elements would be unstable

33. Cosmological constant:

if larger: Universe would expand too quickly to form solar-type stars (see: "Evidence for the Fine-Tuning of the Universe").

Consider also these additional fine-tuning examples:

Ratio of electrons to protons	$1:10^{37}$
Ratio of electromagnetic force to gravity	$1:10^{40}$
Expansion rate	$1:10^{55}$
Mass of Universe	$1:10^{59}$
Cosmological Constant (Lambda)	$1:10^{120}$

In commenting on the difficulty associated with getting the exact ratio of electrons to protons merely "by accident," one astronomer wrote:

> One part in 10^{37} is such an incredibly sensitive balance that it is hard to visualize. The following anal-

ogy might help: Cover the entire North American continent in dimes all the way up to the moon, a height of about 239,000 miles. (In comparison, the money to pay for the U.S. federal government debt would cover one square mile less than two feet deep with dimes.). Next, pile dimes from here to the moon on a billion other continents the same size as North America. Paint one dime red and mix it into the billion of piles of dimes. Blindfold a friend and ask him to pick out one dime. The odds that he will pick the red dime are one in 10^{37} (Ross, 1993, p. 115, parenthetical item in orig.).

And it gets progressively more complicated, as John G. Cramer observed:

A similar problem is raised by the remarkable "flatness" of the universe, the nearly precise balance between expansion energy and gravitational pull, which are within about 15% of perfect balance. Consider the mass of the universe as a cannonball fired upward against gravity at the Big Bang, a cannonball that for the past 8 billion years has been rising ever more slowly against the pull. The extremely large initial kinetic energy has been nearly cancelled by the extremely large gravitational energy debt. The remaining expansion velocity is only a tiny fraction of the initial velocity. The very small remaining expansion kinetic energy and gravitational potential energy are *still* within 15% of one another. To accomplish this the original energy values at one second after the Big Bang must have matched to one part in 10^{15}. **That two independent variables should match to such unimaginably high precision seems unlikely** (1999, italics in orig.; emp. added).

At every turn, there are more examples of the fact that the Universe is "fine-tuned" to such an incredible degree that it becomes impossible to sustain the belief that it "just happened" as the result of (to quote Victor Stenger) "a random quantum fluctuation in a spaceless, timeless void." For example, cosmologists speak of a number known as the "Omega" value. In *Wrinkles of Time*, physicists Smoot and Davidson discussed Omega as follows.

If the density of the mass in the universe is poised precisely at the boundary between the diverging paths to ultimate collapse and indefinite expansion, then the Hubble expansion may be slowed, perhaps coasting to a halt, but never reversed. This happy state of affairs is termed the **critical density**.

The critical density is calculated to be about five millionths of a trillionth of a trillionth (5×10^{-30}) of a gram of matter per cubic centimeter of space, or equivalent to about one hydrogen atom in every cubic meter—a few in a typical room. This sounds vanishingly small, and it is.... If we know the critical density, then we can—in theory—begin to figure out our fate. All we have to do is count up all mass in the universe and compare it to the critical density. **The ratio of the actual density of mass in the universe to the critical density is known, ominously, by the last letter in the Greek alphabet, Omega, Ω.** An Omega of less than 1 leads to an open universe (the big chill), and more than 1 to a closed universe (the big crunch). An Omega of exactly 1 produces a flat universe....

The important thing to remember is that the shape, mass, and fate of the cosmos are inextricably linked; they constitute a single subject, not three. These three aspects come together in, in Omega, the ratio of the actual density to the critical density. The task of measuring the actual density of the universe is extremely challenging, and most measurements produce only approximate figures.... What's the bottom line?... [W]e arrive at an average density of the universe of close to the critical density: Omega is close to 1.... If Omega were well below 1, however, then very few regions would collapse. If Omega were well above 1, then everything would collapse. The closer Omega is to 1, the easier it is to form the structure of the universe that astronomers now observe....

When we learn of the consequences of Omega being anything other than precisely 1, we see how very easily our universe might not have come into existence: The most minute deviation either side of an Omega of 1 consigns our potential universe to oblivion.... There is a long list of physical laws and conditions that, varied slightly,

**would have resulted in a very different universe,
or no universe at all. The Omega-equals-1 requirement is among them** (1993, pp. 158,160,161,
190, emp. added).

The problem, however, is not just that Omega must be so
very exact. A "flat" Universe is one that continues to expand
forever, but at a rate that is so strongly influenced by gravitational forces that the expansion gradually slows down over
billions of years and eventually almost stops. For this to occur, however, the Universe would have to be exactly at critical density. Yet as Roy C. Martin Jr. pointed out in his book,
Astronomy on Trial:

> A critical density, a very, very, very critical density,
> would be required to just balance the expansion with
> gravitation. The trouble is that the required balance
> of forces is so exact, that the chance of it happening
> would have to be something like one in a thousand
> trillions, and no measurements, or mathematics, or
> even theory supports a concept of such exactness. It
> would take an enormous amount of luck for a Flat
> universe to evolve, and it is just about mathemati-
> cally impossible.

> As we said, scientists favor this model, even though
> there is no scientific justification whatsoever for their
> choosing this over any other. Why is this idea popu-
> lar? Well, if you and I were given the choice of a uni-
> verse scheduled for a slow death, one scheduled to
> collapse in a big crunch, or a universe scheduled to
> go on forever, which would we choose? We all, scien-
> tist and not, consider an ongoing Flat universe far
> more palatable. It's merely intuitive, of course, but
> scientists are human also. It should not be missed that
> the Flat, ongoing universe, the one that is almost math-
> ematically impossible, is the closest to an infinitely
> lasting universe that **could not** have been born in a
> Big Bang, and the closest to what we observe! (1999,
> p. 160).

And so, once more science has found itself face-to-face with
yet another inexplicable, finely tuned force of nature that
"somehow" must be explained by blind, random, naturalis-

tic forces. One would think that, after confronting **so many** of these finely tuned forces, scientists finally would admit the obvious. To use the words of evolutionist H.S. Lipson of Great Britain: "I think, however, that we must go further than this and admit that the only acceptable explanation is **creation**"(1980, 31:138, emp. in orig.).

Science is based on observation and reproducibility. But when pressed for the reproducible, empirical data that document their claim of a self-created Universe, scientists and philosophers are at a loss to produce those data. Perhaps this is why Alan Guth, co-developer of the original inflationary Universe theory, lamented: "In the end, I must admit that questions of plausibility are not logically determinable and depend somewhat on intuition" (1988, 11 [2]:76)–which is little more than a fancy way of saying, "I certainly **wish** this were true, but I could not **prove** it to you if my life depended on it." To suggest that the Universe created itself is to posit a self-contradictory position. Sproul addressed this when he wrote:

> For something to bring itself into being it must have the power of being within itself. It must at least have enough causal power to cause its own being. If it derives its being from some other source, then it clearly would not be either self-existent or self-created. It would be, plainly and simply, an effect. Of course, the problem is complicated by the other necessity we've labored so painstakingly to establish: It would have to have the causal power of being before it was. It would have to have the power of being before it had any being with which to exercise that power (1994, p. 180).

The Universe is not eternal. Nor did not create itself from nothing.

Scientifically, the choice is between **matter only** and **more than matter** as the fundamental explanation for the existence and orderliness of the Universe. The difference, therefore, between the evolution model and the creation model is the difference between: (a) **time, chance, and the inherent properties of matter;** or (b) **design, creation, and the ir-**

reducible properties of organization. In fact, when it comes to any particular case, there are again only two scientific explanations for the origin of the order that characterizes the Universe and life in the Universe: either the order was **imposed on** matter, or it **resides within** matter. However, if it is suggested that the order resides within matter, we respond by saying that we certainly have not seen the evidence of such. The creation model not only is plausible, but also is the only one that postulates an adequate cause for the Universe and life in that Universe. The evolution model cannot, and does not. The evidence speaks clearly to the existence of a non-contingent, eternal, self-existent Mind that created this Universe and everything within it.

4

THE LAW OF CAUSE AND EFFECT

Indisputably, the most universal, and most certain, of all scientific laws is the Law of Cause and Effect, or as it is commonly known, the Law of Causality. In science, laws are seen as "reflecting actual regularities in nature" (Hull, 1974, p. 3). So far as historical experience can attest, laws know no exceptions. And this certainly is true of the Law of Causality. This law has been stated in a variety of ways, each of which adequately expresses its ultimate meaning. Kant, in the first edition of *Critique of Pure Reason*, stated that "everything that happens (begins to be) presupposes something which it follows according to a rule." In the second edition, he strengthened that statement by noting that "all changes take place according to the law of connection of cause and effect" (see Meiklejohn, 1878, p. 141). Schopenhauer stated the proposition as: "Nothing happens without a reason why it should happen rather than not happen" (see von Mises, 1968, p. 159). The number of various formulations could be expanded almost indefinitely. But simply put, the Law of Causality states that **every material effect must have an adequate antecedent cause**.

The philosophical/theological implications of this concept—pro and con—have been argued through the years. But after the dust settles, the Law of Causality always remains intact.

There is no question of its acceptance in the world of experimental science or in the ordinary world of personal experience. Many years ago, professor W.T. Stace, in his classic work, *A Critical History of Greek Philosophy*, commented:

> Every student of logic knows that this is the ultimate canon of the sciences, the foundation of them all. If we did not believe the truth of causation, namely, everything which has a beginning has a cause, and that in the same circumstances the same things invariably happen, all the sciences would at once crumble to dust. In every scientific investigation this truth is assumed (1934, p. 6).

The Law of Causality is not just of importance to science. Richard von Mises observed: "We may only add that almost all philosophers regard the law of causality as the most important, the most far-reaching, and the most firmly founded of all principles of epistemology." He then added:

> The law of causality claims that for **every** observable phenomenon (let us call it *B*) there exists a second phenomenon *A*, such that the sentence "*B* follows from *A*" is true.... There can be no doubt that the law of causality in the formulation just stated is in agreement with all our own experiences and with those which come to our knowledge in one way or another.... [We] can also state that in practical life there is hardly a more useful and more reliable rule of behavior than to **assume** of any occurrence that we come to know that some other one preceded it as its cause (1968, p. 160, emp. in orig.).

Richard Taylor, addressing the importance of this basic law of science in *The Encyclopedia of Philosophy*, wrote:

> Nevertheless, it is hardly disputable that the idea of causation is not only indispensable in the common affairs of life but in all applied science as well. Jurisprudence and law would become quite meaningless if men were not entitled to seek the causes of various unwanted events such as violent deaths, fires, and accidents. The same is true in such areas as public health, medicine, military planning, and, indeed, every area of life (1967, p. 57).

SCIENCE AND THE LAW
OF CAUSE AND EFFECT

While the Law of Cause and Effect crosses strictly scientific boundaries and impacts all other disciplines as well, and while the principle of causality has serious theological and/or metaphysical implications in its own right, the scientific implications it presents are among the most serious ever discovered. Obviously, if every material effect has an adequate antecedent cause, and if the Universe is a material effect, then the Universe had a cause. This particular point has not been overlooked by scientists. Robert Jastrow wrote:

> The Universe, and everything that has happened in it since the beginning of time, are a grand effect without a known cause. An effect without a cause? That is not the world of science; it is a world of witchcraft, of wild events and the whims of demons, a medieval world that science has tried to banish. As scientists, what are we to make of this picture? I do not know. I would only like to present the evidence for the statement that the Universe, and man himself, originated in a moment when time began (1977, p. 21).

Effects without adequate causes are unknown. Yet the Universe, says Dr. Jastrow, is a tremendous effect—without any known cause. Centuries of research have taught us much about causes, however. We know, for example, that causes never occur subsequent to the effect. As Taylor has observed:

> Contemporary philosophers...have nevertheless, for the most part, agreed that causes cannot occur after their effects.... [I]t is generally thought to be simply part of the usual meaning of "cause" that a cause is something temporally prior to, or at least not subsequent to, its effect (1967, p. 59).

It is meaningless to speak of a cause following an effect, or of an effect preceding a cause.

We also know, as stated earlier, that the effect never is quantitatively greater than, or qualitatively superior to, the cause. It is this knowledge that is responsible for our formulation of the Law of Causality in these words: "Every material effect must have an **adequate** antecedent cause." The river did not turn muddy because the frog jumped in; the book did not fall from the table because the fly lighted on it; these are not **adequate** causes. For whatever effects we observe, we must postulate adequate causes.

Thus, the Law of Causality has serious implications in every field of endeavor—be it science, metaphysics, or theology. The Universe is here. Some cause prior to the Universe is responsible for its existence. That cause must be greater than, and superior to, the Universe itself. But, as Jastrow has noted, "...the latest astronomical results indicate that at some point in the past the chain of cause and effect terminated abruptly. An important event occurred—the origin of the world—for which there is no known cause or explanation" (1977, p. 27). Of course, when Dr. Jastrow speaks of "no known cause or explanation," he means that there is no known **natural** cause or explanation. Scientists and philosophers alike understand that the Universe must have had a cause. They understand that this cause had to precede the Universe, and be superior to it. Admittedly, there is no natural cause sufficient to explain the origin of matter, and thus the Universe, as Jastrow candidly stated. This presents a very real problem, however, regarding which R.L. Wysong wrote:

> Everyone concludes naturally and comfortably that highly ordered and designed items (machines, houses, etc.) owe existence to a designer. It is unnatural to conclude otherwise. But evolution asks us to break stride from what is natural to believe and then believe in that which is unnatural, unreasonable, and...unbelievable. We are told by some that all of reality—the Universe, life, etc.—is without an initial cause. But, since the Universe operates by cause and effect relationships, how can it be argued from science—which is a study of that very Universe—that the Universe is without an initial cause? Or, if the evolutionist cites a cause, he cites either eternal matter or energy. Then he has suggested a cause far less than the effect. The basis for this departure from what is natural and reasonable to believe is not fact, observation, or experi-

ence but rather unreasonable extrapolations from abstract probabilities, mathematics, and philosophy (1976, p. 412, ellipsis in orig.).

Dr. Wysong presented an interesting historical case to document his point. Some years ago, scientists were called to Great Britain to study, on the Salisbury Plain at Wiltshire, the orderly patterns of concentric rocks and holes at Stonehenge. As studies progressed, it became apparent that these patterns had been designed specifically to allow certain astronomical predictions. The questions of how the rocks were moved into place, how these ancient people were able to construct an astronomical observatory, how the data derived from their studies were used, and many others remain unsolved. But one thing is clear: the **cause** of Stonehenge was intelligent design.

Now, suggested Dr. Wysong, compare Stonehenge (as one television commentary did) to the situation paralleling the origin of life. We study life, observe its various functions, contemplate its complexity (which admittedly defies duplication even by intelligent men with the most advanced methodology and technology)—and what is our conclusion? Theoretically, Stonehenge **might** have been produced by the erosion of a mountain, or by catastrophic natural forces (like tornadoes or hurricanes) working in conjunction with meteorites to produce rock formations and concentric holes. But what practicing scientist (or for that matter, television commentator) ever would entertain seriously such a ridiculous idea? And what person with any common sense would believe such a suggestion?

Yet with the creation of life—the intricate design of which makes Stonehenge look like something a three-year-old child assembled on a Saturday afternoon in the middle of a blinding rainstorm using Mattel building blocks—we are being asked to believe that such can be explained by blind, mindless, accidental, random processes without any intelligent direction whatsoever. It is not surprising that Dr. Wysong should observe, with obvious discomfort, that evolutionists ask us to "break stride with what is natural to believe" in this regard.

No one ever could be convinced that Stonehenge "just happened." That is not an adequate cause. Yet we are expected to accept that life "just happened." Such a conclusion is both unwarranted and unreasonable. The cause is inadequate to produce the effect.

It is this understanding of the implications of the Law of Causality that has led some to attempt to discredit, or refuse to accept, the universal principle of cause and effect. Perhaps the most famous skeptic in this regard was the British empiricist, David Hume, who was renowned for his antagonism to the principle of cause and effect. However, as fervent as Hume was in his criticism, he never went so far as to assert that cause and effect did not exist. He simply felt that it was not empirically verifiable, and stemmed instead from *a priori* considerations. Hume commented in a letter to John Stewart:

> I never asserted so absurd a Proposition as that **anything might arise without a Cause**: I only maintained, that our Certainty of the Falsehood of that Proposition proceeded neither from Intuition nor Demonstration; but from another Source (see Greig, 1932, p. 187, emp. and capital letters in orig.; Craig, 1984, p. 75).

Even so rank an infidel as Hume would not deny cause and effect.

Try as they might, skeptics are unable to circumvent this basic law of science. Arguments other than those raised by Hume have been leveled against it, of course. For example, one such argument insists that the principle must be false because it is inconsistent with itself. The argument goes something like this. The principle of cause and effect says that everything must have a cause. On this concept, it then traces all things back to a First Cause, where it suddenly stops. But how may it consistently do so? Why does the principle that everything needs a cause suddenly cease to be true? Why is it that this so-called First Cause does not likewise need a cause? If everything else needs an explanation, or a cause, why does this First Cause not also need an explanation, or a cause? And if this First Cause does not need an explanation, why, then, do all other things need one?

Two responses may be offered to such a complaint against causality. First, it is impossible–from a logical standpoint–to defend any concept of "infinite regress" that postulates an endless series of effects with no ultimate first cause. Philosophers have argued this point correctly for generations (see Craig, 1979, pp. 47-51; 1984, pp. 75-81). Whatever begins to exist must have a cause. Nothing causeless happens.

Second, the complaint offered by skeptics suggesting that the Law of Causality is inconsistent with itself is not a valid objection against the Law; rather it is an objection to an **incorrect statement** of that Law. If someone were to say, "Everything must have a cause," then the objection might be valid. But this is not what the Law of Causality says. It states that every **material effect** must have an adequate antecedent cause. As John H. Gerstner correctly reasoned:

> Because every effect must have a cause, there must ultimately be one cause that is not an effect but pure cause, or how, indeed, can one explain effects? A cause that is itself an effect would not explain anything but would require another explanation. That, in turn, would require another explanation, and there would be a deadly infinite regress. But the argument has shown that the universe as we know it is an effect and cannot be self-explanatory; it requires something to explain it which is not, like itself, an effect. There must be an uncaused cause. That point stands (1967, p. 53).

Indeed, the point does stand. Science, and common sense, so dictate. Taylor has noted: "If, however, one professes to find no difference between the relation of a cause to its effect, on the one hand, and of an effect to its cause, on the other, he appears to contradict the common sense of mankind, for the difference appears perfectly apparent to most men..." (1967, p. 66). Once again, it is refreshing to see scholars finally appeal to "common sense" or that which is "perfectly apparent to most men." In the case of the Law of Causality, it is "perfectly apparent" that every material effect must have an adequate cause.

Although critics have railed against, and evolutionists have ignored, the Law of the Cause and Effect, it stands unassailed. Its central message remains intact: **Every material effect must have an adequate antecedent cause.** The Universe is here. Life in our magnificent Universe is here. Intelligence is here. Morality is here. What is their ultimate cause? Since the effect never is prior, or superior, to the cause, it stands to reason that the Cause of life must be both antecedent to, and more powerful than, the Universe—a living Intelligence that is Itself of a moral nature. While the evolutionist is forced to concede that the Universe is "an effect without a known cause" (to use Dr. Jastrow's words), the creationist postulates an adequate Cause—a transcendent Creator—that is in keeping with the known facts and the implications accompanying those facts.

5

THE LAW OF BIOGENESIS

In the field of biology, one of the most commonly accepted and widely used laws of science is the Law of Biogenesis. This law was set forth many years ago to dictate what both theory and experimental evidence showed to be true among living organisms—that life comes only from preceding life of its own type or kind. David Kirk observed:

> By the end of the nineteenth century there was general agreement that life cannot arise from the nonliving under conditions that now exist upon our planet. The dictum "All life from preexisting life" became the dogma of modern biology, from which no reasonable man could be expected to dissent (1975, p. 7).

Experiments that ultimately formed the basis of this law were carried out first by such men as Francesco Redi (1688) and Lazarro Spallanzani (1799) in Italy, Louis Pasteur (1860) in France, and Rudolph Virchow (1858) in Germany. It was Virchow who documented that cells do not arise from amorphous matter, but instead come only from preexisting cells. The *Encyclopaedia Britannica* stated concerning Virchow that "His aphorism '*omnis cellula e cellula*' (every cell arises from a pre-existing cell) ranks with Pasteur's '*omne vivum e vivo*' (every living thing arises from a preexisting living thing) among the most revolutionary generalizations of biology" (see Ackerknect, 1973, p. 35).

Through the years, countless thousands of scientists in various disciplines have established the Law of Biogenesis as just that—a scientific law stating that life comes only from preexisting life. Interestingly, the Law of Biogenesis was established firmly in science long before the contrivance of modern evolutionary theories. Also of considerable interest is the fact that students are taught consistently in high school and college biology classes the tremendous impact of, for example, Pasteur's work on the false concept of spontaneous generation (the idea that life arises on its own from nonliving antecedents). Students are given, in great detail, the historical scenario of how Pasteur triumphed over "mythology," providing science with "its finest hour" as he discredited the popular concept of spontaneous generation. Then, with almost the next breath, students are informed by the professor that evolution started via spontaneous` generation.

Abiogenesis, or as it is known more commonly, spontaneous generation, is one of the foundational concepts of evolution. In 1960, when G.A. Kerkut published his famous book, *The Implications of Evolution*, he listed the seven **nonprovable assumptions** upon which evolution is based. At the top of that list was: "The first assumption is that non-living things gave rise to living material, i.e., spontaneous generation occurred" (p. 6). Nobel laureate George Wald of Harvard wrote:

> As for spontaneous generation, it continued to find acceptance until finally disposed of by the work of Louis Pasteur—it is a curious thing that until quite recently professors of biology habitually told this story as part of their introductions of students to biology. They would finish this account glowing with the conviction that they had given a telling demonstration of the overthrow of mystical notion by clean, scientific experimentation. Their students were usually so bemused as to forget to ask the professor how he accounted for the origin of life. This would have been an embarrassing question, because there are only two possibilities: either life arose by spontaneous generation, which the professor had just refuted; or it arose by supernatural creation, which he probably regarded as anti-scientific (1962, p. 187).

Dr. Wald then offered his observations on how to solve this conundrum when he said:

The reasonable view was to believe in spontaneous generation; the only alternative, to believe in a single, primary act of supernatural creation. There is no third alternative.... Most modern biologists, having reviewed with satisfaction the downfall of the spontaneous generation hypothesis, yet unwilling to accept the alternative belief in special creation, are left with nothing. I think a scientist has no choice but to approach the origin of life through a hypothesis of spontaneous generation. What the controversy reviewed above showed to be untenable is only the belief that living organisms arise spontaneously under present conditions. We have now to face a somewhat different problem: how organisms may have arisen spontaneously under different conditions in some former period, granted that they do so no longer.

To make an organism demands the right substances in the right proportions and in the right arrangement. We do not think that anything more is needed—but that is problem enough. One has only to contemplate the magnitude of this task to concede that the spontaneous generation of a living organism is impossible. Yet here we are, as a result, I believe, of spontaneous generation (1979, pp. 289-291).

Notice several things regarding Dr. Wald's statements. First, he admitted to no third alternative. **Either** spontaneous generation (chemical evolution) is true, **or** creation occurred. Second, he granted that spontaneous generation is **not occurring now**. Third, he felt, however, that it **must have occurred** in the distant past. Dr. Wald, of course, was correct when he stated that there are only two choices, and that spontaneous generation is not occurring now. He also was correct in his observation that students often forget to ask their professors how, if spontaneous generation has been discredited, evolution could have gotten started in the first place.

However, while these important points may have escaped some students, they have not been lost on evolutionists, who confess to having some difficulty with such problems. For example, Jastrow has written:

At present, science has no satisfactory answer to the question of the origin of life on the earth. Perhaps the appearance of life on the earth is a miracle. Scientists are reluctant to accept that view, but their choices are limited; **either** life was created on the earth by the will of a being outside the grasp of scientific understanding, **or** it evolved on our planet spontaneously, through chemical reactions occurring in nonliving matter lying on the surface of the planet. The first theory places the question of the origin of life beyond the reach of scientific inquiry. It is a statement of faith in the power of a Supreme Being not subject to the laws of science. The second theory is also an act of faith. The act of faith consists in assuming that the scientific view of the origin of life is correct, without having concrete evidence to support that belief (1977, pp. 62-63, emp. in orig.).

Elsewhere in the same book from which the above quotation was taken, Dr. Jastrow remarked:

> According to this story, every tree, every blade of grass, and every creature in the sea and on the land evolved out of one parent strand of molecular matter drifting lazily in a warm pool. What concrete evidence supports that remarkable theory of the origin of life? There is none (1977, p. 60).

That, of course, is a rather startling admission. Apparently evolutionists continue to believe in spontaneous generation, in spite of the fact that there is no good evidence for it.

In their popular high school biology textbook, *Life: An Introduction to Biology*, Simpson and Beck stated: "...there is no serious doubt that biogenesis is the rule, that life comes only from other life, that a cell, the unit of life, is **always and exclusively** the product or offspring of another cell" (1965, p. 144, emp. added). Martin A. Moe, writing in *Science Digest*, expressed it like this:

> A century of sensational discoveries in the biological sciences has taught us that **life arises only from life**, that the nucleus governs the cell through the molecular mechanisms of deoxyribonucleic acid (DNA) and that the amount of DNA and its structure determine not only the nature of the species but also the characteristics of individuals (1981, 89[11]:36, emp. added).

The late evolutionist Loren Eiseley once stated that in postulating the idea of spontaneous generation, science had "created a mythology of its own" (1957, pp. 201-202). One wonders how much evidence against something there would have to be before it would be discarded? There is one nice thing about having no evidence, however. Richard Dickerson, writing in *Scientific American* under the heading of "Chemical Evolution and the Origin of Life," remarked that we have "no laboratory models: hence one can speculate endlessly, unfettered by inconvenient facts" (1978, p. 85). And, as Dr. Dickerson admitted: "We can only imagine what probably existed, and our imagination so far has not been very helpful" (p. 86).

It is easy, after reviewing the literature on spontaneous generation/chemical evolution, to see how terribly weak the case is for such a scenario. Green and Goldberger hardly could have put it more bluntly when they wrote:

> There is one step [in evolution–BT] that far outweighs the others in enormity: the step from macromolecules to cells. All the other steps can be accounted for on theoretical grounds–if not correctly, at least elegantly. However, the macromolecule to cell transition is a jump of fantastic dimensions, which lies beyond the range of testable hypothesis. In this area, all is conjecture. The available facts do not provide a basis for postulation that cells arose on this planet. This is not to say that some para-physical forces were not at work. We simply wish to point out that there is **no scientific evidence** (1967, pp. 406-407, emp. added).

Hoyle and Wickramasinghe, in their popular text, *Lifecloud*, concluded:

> It is doubtful that anything like the conditions which were simulated in the laboratory existed at all on a primitive Earth, or occurred for long enough times and over sufficiently extended regions of the Earth's surface to produce large enough local concentrations of the biochemicals required for the start of life. In accepting the "primeval soup theory" of the origin of life scientists have replaced religious mysteries which shrouded this question with equally mysterious scientific dogmas. The implied scientific dogmas are just as inaccessible to the empirical approach (1978, p. 26).

Thirteen years later, writing under the intriguing title, "Where Microbes Boldly Went," Hoyle and Wickramasinghe lamented in *New Scientist:*

> Precious little in the way of biochemical evolution could have happened on the Earth. It is easy to show that the two thousand or so enzymes that span the whole of life could not have evolved on the Earth. If one counts the number of trial assemblies of amino acids that are needed to give rise to the enzymes, the probability of their discovery by random shufflings turns out to be less than 1 in $10^{40,000}$ (1991, 91:415).

Sir Francis Crick, co-discoverer of the structure of the DNA molecule, agreed when he wrote a decade earlier:

> If a particular amino acid sequence was selected by chance, how rare an event would this be?
>
> This is an easy exercise in combinatorials. Suppose the chain is about two hundred amino acids long; this is, if anything rather less than the average length of proteins of all types. Since we have just twenty possibilities at each place, the number of possibilities is twenty multiplied by itself some two hundred times. This is conveniently written 20^{200} and is approximately equal to 10^{260}, that is, a one followed by 260 zeros.
>
> ...Moreover, we have only considered a polypeptide chain of rather modest length. Had we considered longer ones as well, the figure would have been even more immense.... The great majority of sequences can never have been synthesized at all, at any time....
>
> **An honest man, armed with all the knowledge available to us now, could only state that in some sense, the origin of life appears at the moment to be almost a miracle, so many are the conditions which would have had to have been satisfied to get it going** (1981, pp. 51-52,88, emp. added).

Four years later, evolutionist Andrew Scott authored an article in *New Scientist* on the origin of life titled "Update on Genesis," in which he observed:

> Take some matter, heat while stirring, and wait. That is the modern version of Genesis. The "fundamental" forces of gravity, electromagnetism and the strong

and weak nuclear forces are presumed to have done the rest.... But how much of this neat tale is firmly established, and how much remains hopeful speculation? In truth, the mechanism of almost every major step, from chemical precursors up to the first recognizable cells, is the subject of either controversy or complete bewilderment.

We are grappling with a classic "chicken and egg" dilemma. Nucleic acids are required to make proteins, whereas proteins are needed to make nucleic acids and also to allow them to direct the process of protein manufacture itself.

The emergence of the gene-protein link, an absolutely vital stage on the way up from lifeless atoms to ourselves, is still shrouded in almost complete mystery.... We still know very little about how our genesis came about, and to provide a more satisfactory account than we have at present remains one of science's great challenges (1985, 106:30-33).

In their text, *The Mystery of Life's Origin*, which is an in-depth review and refutation of experiments on chemical evolution, Thaxton, Bradley, and Olsen stated:

Chemical evolution is broadly regarded as a highly plausible scenario for imagining how life on earth might have begun. It has received support from many competent theorists and experimentalists. Ideas of chemical evolution have been modified and refined considerably through their capable efforts. Many of the findings of these works, however, have not supported the scenario of chemical evolution. In fact, what has emerged over the last three decades, as we have shown in the present critical analysis, is an alternative scenario which is characterized by destruction, and not the synthesis of life.

This alternative scheme envisions a primitive earth with an oxidizing atmosphere. A growing body of evidence supports the view that substantial quantities of molecular oxygen existed very early in earth history before life appeared. If the early atmosphere was strongly oxidizing...then no chemical evolution ever occurred. Even if the primitive atmosphere was reducing or only mildly oxidizing, then degradative processes predominated over synthesis.... The prebiotic chemical soup, presumably a worldwide phenomenon, left no known trace in the geological record.

...There does not seem to be any physical basis for the widespread assumption implicit in the idea that an open system is a sufficient explanation for the complexity of life. As we have previously noted, there is neither a theoretical nor an experimental basis for this hypothesis. There is no hint in our experience of any mechanistic means of supplying the necessary configurational entropy work....

...Notice, however, that the sharp edge of this critique is not what we **do not** know, but what we **do** know. Many facts have come to light in the past three decades of experimental inquiry into life's beginning. With each passing year the criticism has gotten stronger. The advance of science itself is what is challenging the notion that life arose on earth by spontaneous (in a thermodynamic sense) chemical reactions.

...A major conclusion to be drawn from this work is that the undirected flow of energy through a primordial atmosphere and ocean is at present a woefully inadequate explanation for the incredible complexity associated with even simple living systems, and is probably wrong (1984, pp. 182,183,185,186, emp. in orig.).

As these authors have correctly noted, regardless of the type of atmosphere on the primitive Earth (reducing or oxidizing), the singular problem of the tremendously complex information system that somehow must be acquired by living organisms has not been solved. In his 1999 book, *Biogenesis: Theories of Life's Origins*, Noam Lahav admitted:

Thus, by challenging the assumption of a reducing atmosphere, we challenge the very existence of the "prebiotic soup," with its richness of biologically important organic compounds. Moreover, so far, no geochemical evidence for the existence of a prebiotic soup has been published. Indeed, a number of scientists have challenged the prebiotic soup concept, noting that even if it existed, the concentration of organic building blocks in it would have been too small to be meaningful for prebiotic evolution (pp. 138-139).

Evolutionist Douglas Hofstadter remarked:

> A natural and fundamental question to ask on learning of these incredibly interlocking pieces of software and hardware is: "How did they ever get started in the first place?" It is truly a baffling thing. One has to imagine some sort of bootstrap process occurring, somewhat like that which is used in the development of new computer language—but a bootstrap from simple molecules to entire cells is almost beyond one's power to imagine. There are various theories on the origin of life. They all run aground on this most central of all central questions: "How did the Genetic Code, along with the mechanisms for its translation (ribosomes and RNA molecules) originate?" For the moment, we will have to content ourselves with a sense of wonder and awe, rather than with an answer (1980, p. 548).

Leslie Orgel, one of the "heavyweights" in origin-of-life studies, similarly admitted:

> We do not yet understand even the general features of the origin of the genetic code.... The origin of the genetic code is the most baffling aspect of the problem of the origins of life, and a major conceptual or experimental breakthrough may be needed before we can make any substantial progress (1982, p. 151).

Writing in *Nature* on "The Genesis Code by Numbers," evolutionist John Maddox commented:

> It was already clear that the genetic code is not merely an abstraction but the embodiment of life's mechanisms; the consecutive triplets of nucleotides in DNA (called codons) are inherited but they also guide the construction of proteins. So it is disappointing that the origin of the genetic code is still as obscure as the origin of life itself (1994, 367:111).

Just three years earlier, John Horgan authored an article for *Scientific American* titled "In the Beginning," in which he wrote:

> DNA cannot do its work, including forming more DNA, without the help of catalytic proteins, or enzymes. In short, proteins cannot form without DNA, but neither can DNA form without proteins.

But as researchers continue to examine the RNA-world concept closely, more problems emerge. How did RNA arise initially? RNA and its components are difficult to synthesize in a laboratory under the best of conditions, much less under plausible prebiotic ones (1991, 264:119).

In their biology textbook, *The New Biology*, Robert Augros and George Stanciu asked:

> What cause is responsible for the origin of the genetic code and directs it to produce animal and plant species? It cannot be matter because of itself matter has no inclination to these forms, any more than it has to the form Poseidon or the form of a microchip or any other artifact. **There must be a cause apart from matter that is able to shape and direct matter.** Is there anything in our experience like this? Yes, there is: our own minds. The statue's form originates in the mind of the artist, who then subsequently shapes matter, in the appropriate way.... **For the same reasons there must be a mind that directs and shapes matter in organic forms** (1987, p. 191, emp. added).

Creationists are not shocked by such admissions. In spite of all the hullabaloo surrounding origin-of-life experiments, no one has yet "created life," or even come close. In fact, laboratory experiments have not even remotely approached the synthesis of life from nonlife, and the extremely limited results attained thus far have depended upon artificially imposed conditions that were extremely improbable. **In nature, we have not documented a single case of spontaneous generation/chemical evolution.** Cows give rise to cows, birds to birds, tulips to tulips, corn to corn, and so on.

In recent years, however, some evolutionists have suggested that the Law of Biogenesis is not a "law" at all, but only a "principle" or "theory" or "dictum." This new nomenclature is being suggested by evolutionists because they have come to a stark realization of the implications of the Law of Biogenesis—not because contradictions or exceptions to the law have been discovered. In nineteenth-century science texts, biogenesis was spoken of as a **law**. Of late, however, that term has been replaced by new words that are intended to "soften" the force of biogenesis upon evolutionary concepts. A rose by any other name, however, is still a rose. And there can be no doubt that biogenesis most certainly reflects (to use Dr. Hull's words) "an actual regularity in nature," since there never has been even a single documented case of spontaneous generation.

Still, some modern-day evolutionists prefer to use a different term when speaking of biogenesis. Under the heading of "Biogenesis, Principle of," one well-known biology dictionary offered the following definition: "The biological **rule** that a living thing can originate only from a parent or parents on the whole similar to itself. It denies spontaneous generation..." (see Abercrombie, et al., 1961, p. 33, emp. added). Others have followed suit. Simpson and Beck, in their text quoted above, stated: "We take biogenesis as a fundamental **principle** of reproduction from the experimental evidence and also from theoretical considerations" (1965, p. 144, emp. added). Wysong, in *The Creation-Evolution Controversy*, lamented this trend.

> The creationist is quick to remind evolutionists that biopoiesis [or abiogenesis—BT] and evolution describe events that stand in stark naked contradiction to an established law. The law of biogenesis says life arises only from preexisting life, biopoiesis says life sprang from dead chemicals; evolution states that life forms give rise to new, improved and different life forms, the law of biogenesis says that kinds only reproduce their own kinds. Evolutionists are not oblivious to this law. They simply question it. They say that spontaneous generation was disproved under the conditions of the experimental models of Pasteur, Redi, and Spallanzani. This, they contend, does not preclude the spontaneous formation of life under different conditions. To this, the creationist replies that even given the artificial conditions and intelligent maneuverings of biopoiesis experiments, life has still not "spontaneously generated." ...Until such a time as life is observed to spontaneously generate, the creationist insists the law of biogenesis stands!... How can biogenesis be termed any less than a law? (1976, pp. 182,184,185).

Moore and Slusher, in their textbook, *Biology: A Search for Order in Complexity*, wrote: "Historically the point of view that **life comes only from life** has been so well established through the facts revealed by experiment that it is called the Law of Biogenesis." In an accompanying footnote, the authors went on to state:

> Some philosophers call this a **principle** instead of a law, but this is a matter of definition, and definitions are arbitrary. Some scientists call this a **superlaw**, or a law about laws. Regardless of terminology, biogenesis has the highest rank in these levels of generalization (1974, p. 74, emp. in orig.).

Indeed, as Dr. Kirk (quoted above) noted, the dictum "became the dogma of modern biology, from which no reasonable man could be expected to dissent."

Furthermore, it is of interest to turn to the scientific dictionaries and observe the definition of the word "principle" that is being used so often in the current controversy. The *McGraw-Hill Dictionary of Scientific and Technical Terms*, an industry standard, defines principle as, "a scientific **law** which is highly general or fundamental, and from which other laws are derived" (see Lapedes, 1978, p. 1268). The reason that some scientists call biogenesis a **superlaw** has to do with the fact that at times other laws are derived from it (the laws of Mendelian genetics hardly could operate without the fundamental "principle" of biogenesis being correct). If a principle is defined as a law, and biogenesis is spoken of as the "principle of biogenesis," what more shall we say? As Kirk himself noted: "The more broadly encompassing paradigms—those from which the largest and most diverse blocks of biological information may be related in orderly fashion—are sometimes called 'principles' of biology" (1975, p. 14).

In other areas of science besides biology, it is common to hear scientists speak of well-established and readily recognized laws as "principles." Reference often is made to the "principles" of thermodynamics or the "principle" of gravity instead of the "laws" of thermodynamics or the "law" of gravity. Yet no one calls into question these basic and fundamental laws of science. Even in biology we use such terminology (e.g., we speak of the "principles" of Mendelian genetics), without having anyone question the basic nature of the laws of science under discussion.

Why, then, in regard to biogenesis, is it suggested that the term "law" no longer applies? It did in the nineteenth century. Has it been disproved? No. Every shred of scientific evidence still supports the concept that life arises only from pre-existing life. Is biogenesis no longer an "actual regularity in nature"? On the contrary, all of the scientific information we possess shows that it is just that—an actual regularity in nature (recall Dr. Simpson's statement that "there is no serious doubt that biogenesis is the rule, that life comes only from other life...").

Has biogenesis somehow ceased being experimentally reproducible? Hardly. Why, then, do evolutionists insist that biogenesis no longer be referred to as a law? The answer, of course, is obvious. If evolutionists accept biogenesis as a scientific law—i.e., an actual regularity in nature—evolution never could get started. Acknowledging the **Law** of Biogenesis would represent the complete undoing of evolutionary theory from the ground floor up. Thus, some modern-day evolutionists have scoured the dictionary to find another word ("rule," "principle," "dictum," etc.) besides **law** to attach to biogenesis. Regardless of their efforts, one thing is certain: the "dogma of modern biology, from which no reasonable man could be expected to dissent," is still biogenesis. J.W.N. Sullivan, brilliant scientist of the past, penned these words, which are as applicable today as the day he wrote them.

> The beginning of the evolutionary process raises a question which is yet unanswerable. What was the origin of life on this planet? Until fairly recent times there was a pretty general belief in the occurrence of "spontaneous generation." ...But careful experiments, notably those of Pasteur, showed that this conclusion was due to imperfect observation, and **it became an accepted doctrine that life never arises except**

from life. So far as the actual evidence goes, this is still the only possible conclusion. But since it is a conclusion that seems to lead back to some supernatural creative act, it is a conclusion that scientific men find very difficult of acceptance (1933, p. 94, emp. added).

6

THE LAWS OF GENETICS

One of the newest, and certainly one of the most exciting, sciences is that of genetics. After all, every living thing—plant, animal, and human—is a storehouse of genetic information and therefore a potential "laboratory" full of scientific knowledge. Studies have shown that the hereditary information found within the nucleus of the living cell is placed there in a chemical "code," and that this code is universal in nature. Regardless of their respective views on origins, all scientists acknowledge this. Evolutionist Richard Dawkins stated: "The genetic code is universal.... The complete word-for-word universality of the genetic dictionary is, for the taxonomist, too much of a good thing" (1986, p. 270). Creationist Darrel Kautz agreed: "It is recognized by molecular biologists that the genetic code is universal, irrespective of how different living things are in their external appearances" (1988, p. 44).

However, it is not simply the fact that the genetic code is universal in nature that makes its study so appealing. The function of this code is equally intriguing. A.E. Wilder-Smith, the late, eminent scientist from the United Nations, observed:

> The construction and metabolism of a cell are thus dependent upon its internal "handwriting" in the genetic code. Everything, even life itself, is regulated from a biological viewpoint by the information contained in this genetic code. All syntheses are directed by this information (1976, p. 254).

Since all living things are storehouses of genetic information (i.e., within the genetic code), and since it is this code that regulates life and directs its synthesis, the importance of the study of this information code hardly can be overstated. Renowned British geneticist, E.B. Ford, in his book, *Understanding Genetics*, provided an insightful summary in this regard:

> It may seem a platitude to say that the offspring of buttercups, sparrows and human beings are buttercups, sparrows and human beings.... What then keeps them, and indeed living things in general, "on the right lines"? Why are there not pairs of sparrows, for instance, that beget robins, or some other species of bird: why indeed birds at all? Something must be handed on from parent to offspring which ensures conformity, not complete but in a high degree, and prevents such extreme departures. What is it, how does it work, what rules does it obey and why does it apparently allow only limited variation? Genetics is the science that endeavours to answer these questions, and much else besides. It is the study of organic inheritance and variation, if we must use more formal language (1979, p. 13).

We know, of course, that sparrows, buttercups, and human beings give rise only to sparrows, buttercups, and human beings. But we know this today because of our in-depth knowledge of genetics—the study of inheritance. However, it has not always been so. The history of how we stumbled upon this knowledge, and thus this new science, provides an interesting, and profitable, case study.

Various writers have chronicled early attempts at hybridization, selection, etc. (see Suzuki and Knudtson, 1989, pp. 32-35). But it is agreed unanimously that the true origin of the science we call genetics had its origin in 1865 as the result of studies performed by an Augustinian monk, Gregor Mendel (1865). In 1857, Mendel began a series of experiments in the garden of the abbey in Brünn, Austria, using edible peas (*Pisum sativum*). For eight years he worked with these peas. The story of Mendel's research is beyond the purview of this book. However, it has been recorded by numerous writers (see Asimov,

1972, pp. 366-368; Gardner, 1972, pp. 401-403; Edey and Johanson, 1989, pp. 108-122; Suzuki and Knudtson, 1989, pp. 35-38; Henig, 2000).

Mendel's accomplishments hardly can be overstated. Richard von Mises observed that Mendel's work "...plays in genetics a role comparable to that of Newton's laws in mechanics" (1968, p. 243). Edey and Johanson echoed that same sentiment:

> Mendel was certain that his hypothesis was correct: hereditary traits of living things come in separate packages; they do not blend; they behave according to simple mathematical laws; some are dominant and "show," while others are recessive and lie "hidden" unless present in the pure state. This was a momentous insight. It became the keystone for the great edifice of genetic knowledge that would be erected in the following century (1989, p. 114).

Davis and Kenyon (1989, p. 60) have summarized what now are known as "Mendel's laws."

1. The inheritance of traits is determined by (what were later termed) genes that act more like individual physical particles than like fluid.

2. Genes come in pairs for each trait, and the genes of a pair may be alike or different.

3. When genes controlling a particular trait are different, the effect of one is observed (dominant) in the offspring, while the other one remains hidden (recessive).

4. In gametes (eggs and sperm) only one gene of each pair is present. At fertilization gametes unite randomly, which results in a predictable ratio of traits among offspring.

5. The genes controlling a particular trait are separated during gamete-formation; each gamete carries only one gene of each pair.

6. When two pairs of traits are studied in the same cross, they are found to sort independently of each other.

In 1866, Mendel's work was published in the *Transactions of the Natural History Society of Brünn*. For thirty-five years, Mendel's work sat on library shelves, unknown to all but a few, and causing no great interest among them. Then, in 1900, three scientists, working independently of one another, rediscovered Mendel's material. Hugo de Vries (a Dutchman), Karl Correns (a German), and Erich Tschermak (an Austrian) simultaneously read Mendel's works and published their own papers on similar matters, each acknowledging Mendel's contribution. De Vries is credited with discovering genetic mutations (changes in the genes and/or chromosomes, producing offspring unlike the parents).

Gregor Mendel died in 1884, never realizing that eventually he would become known as the "Father of Genetics" (see Considine, 1976, p. 1155). Many scientists since have added to the knowledge he provided in regard to this important new science. For example, in 1902, German embryologist Theodor Boveri, and in 1904, American cytologist W.S. Sutton, building on the work of another German embryologist, Wilhelm Roux, documented that what Mendel had referred to as *Anlagen* (genes?) were distributed throughout the body in the nucleus of every cell in sausage-shaped bodies that Roux called "chromosomes" (from the Greek meaning "color body," because early geneticists had to stain them with brightly colored dyes in order to view them under a microscope). In 1906, at a meeting of the Royal Horticultural Society, English biologist William Bateson offered the term "genetics" as the name for this new science. Finally, Mendel's efforts were receiving the recognition they so richly deserved.

The effort to locate a gene, determine what it does, and discover how it functions was launched in 1906 when American scientist Thomas H. Morgan began his famous studies on the chromosomes of fruit flies. That same year, at a meeting of the Royal Horticultural Society, English biologist William Bateson suggested the term "genetics" as the name for this new science (see Asimov, 1972, p. 516). In 1908, Morgan identified "invisible heredity units" (that later would come to be known as genes) as being associated with portions of chromosomes. Then, in 1909, Danish botanist Wilhelm Johannsen coined the term "gene" (from the Greek for "giving birth to") as the name for these "heredity units"—a term still in use today (see Bishop and Waldholz, 1999, p. 23). [Johannsen also coined the two terms "genotype" and "phenotype" to describe an individual's inner genetic make-up, and the outward expression of that make-up, respectively.]

The physical location of the gene, therefore, has been known only since the beginning of this century. Shortly thereafter, it became clear that almost every biochemical characteristic in all living creatures was determined by genes. In 1911, scientists produced the first chromosome maps. In the 1940s, O.T. Avery showed that traits could be passed from one bacterium to another by a chemical known as DNA (see Avery, et al., 1944, 79:137-158). The eminent taxonomist of Harvard, Ernst Mayr, wrote concerning this event: "A new era in developmental genetics was opened when Avery demonstrated that DNA was the carrier of the genetic information" (1997, p. 166). By 1941, two Americans, George Beadle and Edward Tatum, had discovered that the genes' function was to produce proteins—which serve both as structural components of all living matter and as enzymes that assist in the infinite variety of chemical reactions that make life possible. Yet, as Bishop and Waldholz noted:

> Despite these remarkable discoveries, the exact nature of the genes remained a mystery. No one knew what a gene looked like, how it worked, or how the cell managed to replicate its genes in order to pass a complement on to its offspring. By the 1940s, however, a series of discoveries began suggesting that the genes were composed of an acid found in the nuclei of cells. This nucleic acid was rich in a sugar called deoxyribose and hence was known as deoxyribonucleic acid, or DNA (1999, p. 23).

The still-new science of genetics was advanced greatly by the discovery, in 1953, of the chemical code within cells that provides the genetic instructions. It was in that year that James D. Watson of the United States, and Francis H.C. Crick of Great Britain, published their landmark paper about the composition and helical structure of DNA (1953, 171:737-738). Nine years later, in 1962, they were awarded the Nobel Prize in Medicine or Physiology for their stellar achievement in elucidating the structure of DNA (a subject about which I will have more to say later in this chapter). Thaxton, Bradley, and Olsen, in their book, *The Mystery of Life's Origin*, remarked:

> According to their now-famous model, hereditary information is transmitted from one generation to the next by means of a simple code resident in the specific sequence of certain constituents of the DNA molecule.... The breakthrough by Crick and Watson was their discovery of the specific key to life's diversity. It was the extraordinarily complex yet orderly architecture of the DNA molecule. They had discovered that there is in fact a code inscribed in this "coil of life," bringing a major advance in our understanding of life's remarkable structure (1984, p. 1).

Thus, the DNA contains the information that allows proteins to be manufactured, and proteins control cell growth and function, which ultimately are responsible for each organism. The genetic code, as found within the DNA molecule, is vital to life as we know it.

A LOOK AT THE INNER WORKINGS OF THE CELL

As scientists have studied what Dr. Ford (quoted earlier) referred to as "organic inheritance and variation," we have come to realize that the basic unit of life is the cell. Genes, chromosomes, nucleic acids, and the chemicals that compose them are found within the cells of every living organism on Earth. It is quite appropriate, therefore, that an investigation into matters such as those being discussed here should begin with an examination of the structure and nature of the cell.

Anatomist Ernst Haeckel, Charles Darwin's chief supporter in Germany in the mid-nineteenth century, once summarized his personal feelings about the "simple" nature of the cell when he wrote that it contained merely "homogeneous globules of plasm" that were composed chiefly of carbon with an admixture of hydrogen, nitrogen, and sulfur. These component parts properly united produce the soul and body of the animated world, and suitably nursed became man. With this single argument the mystery of the universe is explained, the Deity annulled, and a new era of infinite knowledge ushered in (1905, p. 111).

Voilà! As easy as that, simple "homogeneous globules of plasm" nursed man into existence, animated his body, dispelled the necessity of a Creator, and ushered in a new era of "infinite knowledge." In the end, however, Haeckel's simplistic, naturalistic concept turned out to be little more than wishful thinking. As Lester and Hefley put it:

> We once thought that the cell, the basic unit of life, was a simple bag of protoplasm. Then we learned that each cell in any life form is a teeming micro-universe of compartments, structures, and chemical agents— and each human being has billions of cells... (1998, pp. 30-31).

Billions of cells indeed! In the section he authored on the topic of "life" for the *Encyclopaedia Britannica*, the late astronomer Carl Sagan observed that a single human being is composed of what he referred to as an "ambulatory collection of 10^{14} cells" (1997, 22:965). He then noted: "The information content of a simple cell has been established as around 10^{12} bits, comparable to about a hundred million pages of the *Encyclopaedia Britannica*" (22:966). Evolutionist Richard Dawkins acknowledged that the cell's nucleus "contains a digitally coded database larger, in information content, than all 30 volumes of the *Encyclopaedia Britannica* put together. And this figure is for **each** cell, not all the cells of a body put together" (1986, pp. 17-18, emp. in orig.). Dr. Sagan estimated that if a person were to count every letter in every word in every book of the world's largest library (approximately 10 million volumes), the total number of letters would be 10^{12}, which suggests that the "simple cell" contains the information equivalent of the world's largest library (1974, 10:894)! Stephen C. Meyer suggested:

Since the late 1950s advances in molecular biology and biochemistry have revolutionized our understanding of the miniature world within the cell. Modern molecular biology has revealed that living cells—the fundamental units of life—possess the ability to store, edit and transmit information and to use information to regulate their most fundamental metabolic processes. Far from characterizing cells as simple "homogeneous globules of plasm," as did Ernst Haeckel and other nineteenth-century biologists, modern biologists now describe cells as, among other things, "distributive real-time computers" and complex information processing systems (1998, pp. 113-114).

So much for the "simple" cell being a little lump of albuminous combination of carbon, as Haeckel once put it.

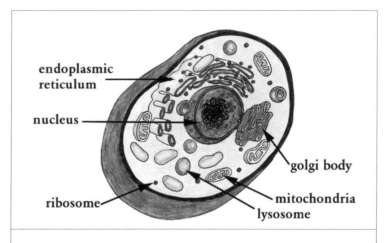

Figure 1 – Simplified representation of a typical eukaryotic cell as rendered by Gabriela Weaver of Colorado University at Denver. Used by permission of Dr. Weaver and The Food Zone [http://Kauai.cudenver.edu:3010/]

Cells are filled with a variety of organelles such as ribosomes (which aid in protein production), Golgi bodies (which package proteins), the endoplasmic reticulum (the transport system of the cell), mitochondria (which manufacture energy), vacuoles (which aid in intracellular cleaning processes), etc. [NOTE: A glossary of terms has been provided in the Appendix of this book for those who may be unfamiliar with the biological/genetic descriptions employed here.] Furthermore, cells are absolute marvels of design when it comes to reproducing themselves. Cellular reproduction consists of at least two important functions—duplication of the cell's complement of genetic material, and cleavage of the cell's cytoplasmic matrix into two distinct-yet-separate parts. However, not all cells reproduce in the same manner.

Speaking in broad terms, there are two basic types of cells found in organisms that procreate sexually. First, there are somatic (body) cells that contain a full complement (the diploid number) of genes. Second, there are germ (egg and sperm) cells that contain half the complement (the haploid number) of genes. Likely, the reason that germ cells (gametes) contain only half the normal genetic content is fairly obvious. Since the genetic material in the two gametes is combined during procreation in order to form a zygote (which will develop first into an embryo, then into a fetus, and eventually into the neonate), in order to ensure that the zygote has the normal, standard chromosome number the gametes always must contain exactly half that necessary number. As Weisz and Keogh explained in their widely used textbook, *Elements of Biology*:

> One consequence of every sexual process is that a zygote formed from two gametes possesses twice the number of chromosomes present in a single gamete. An adult organism developing from such a zygote would consist of cells having a doubled chromosome number. If the next generation is again produced sexually, the chromosome number would quadruple, and this process of progressive doubling would continue indefinitely through successive generations. Such events do not happen, and chromosome numbers do stay constant from one life cycle to the next (1977, p. 331).

Why is it, though, that chromosome numbers "do stay constant from one life cycle to the next?" The answer, of course, has to do with the two different types of cellular division. All somatic cells reproduce by the process known as **mitosis**. Most cells in sexually reproducing organisms possess a nucleus that contains a preset number of chromosomes. In mitosis, cell division is "a mathematically precise doubling of the chromosomes and their genes. The two chromosome sets so produced then become separated and become part of two newly formed nuclei" so that "the net result of cell division is the formation of two cells that match each other and the parent cell precisely in their gene contents and that contain approximately equal amounts and types of all other components (Weisz and Keogh, pp. 322). Thus, mitosis carefully maintains a constant diploid chromosome number during cellular division. For example, in human somatic cells, there are 46 chromosomes. During mitosis, two new "daughter" cells are produced from the original "parent" cell, each of which then contains 46 chromosomes.

Germ cells, on the other hand, reproduce by a process known as **meiosis**. During this type of cellular division, the diploid chromosome number is halved ("meiosis" derives from the Greek meaning to split or divide). So, to use the example of the human, the diploid chromosome complement of 46 is reduced to 23 in each one of the newly formed cells. As Weisz and Keogh observed:

> Meiosis occurs in every life cycle that includes a sexual process—in other words, more or less universally.... It is the function of meiosis to counteract the chromosome-doubling effect of fertilization by reducing a doubled chromosome number to half. The unreduced doubled chromosome number, before meiosis, is called the **diploid** number; the reduced number, after meiosis, is the **haploid** number (p. 331, emp. in orig.).

In his book, *The Panda's Thumb*, evolutionist Stephen Jay Gould discussed the marvel of meiosis.

> Meiosis, the splitting of chromosome pairs in the formation of sex cells, represents one of the great triumphs of good engineering in biology. Sexual reproduction cannot work unless eggs and sperm each con-

tain precisely half the genetic information of normal body cells. The union of two halves by fertilization restores the full amount of genetic information.... This halving, or "reduction division," occurs during meiosis when the chromosomes line up in pairs and pull apart, one member of each pair moving to each of the sex cells. Our admiration for the precision of meiosis can only increase when we learn that cells of some ferns contain more than 600 pairs of chromosomes and that, in most cases, meiosis splits each pair without error (1980a, p. 160).

And it is not just meiosis that works in most instances without error. Evolutionist John Gribbin admitted, for example, that "...once a fertilized, single human cell begins to develop, the original plans are **faithfully copied** each time the cell divides (a process called mitosis) so that every one of the thousand million million cells in my body, and in yours, contains a perfect replica of the original plans for the whole body" (1981, p. 193, parenthetical comment in orig., emp. added).

Regarding the "perfect replica" produced in cellular division, information scientist Werner Gitt remarked:

> The DNA is structured in such a way that it can be replicated every time a cell divides in two. Each of the two daughter cells has to have identically the same genetic information after the division and copying process. This replication is so precise that it can be compared to 280 clerks copying the entire Bible sequentially each one from the previous one, with at most a single letter being transposed erroneously in the entire copying process.... One cell division lasts from 20 to 80 minutes, and during this time the entire molecular library, equivalent to one thousand books, is copied correctly (1997, p. 90).

But as great an engineering triumph as cellular division and reproduction are, they represent only a small part of the story regarding the marvelous design built into each living cell. Since all living things are storehouses of genetic information (i.e., within the genetic code), and since it is this cellular code that regulates life and directs its synthesis, the importance of the study of this code hardly can be overstated.

DNA, GENES, AND CHROMOSOMES

In most organisms, the primary genetic material is DNA [although some viruses, primarily retroviruses, contain only RNA (see Nicholl, 1994, pp. 9-10; Ridley, 1999, p. 9).] What is DNA, and how does it work? [It is not my intention here to present an extremely in-depth examination of the inner workings of the DNA molecule. Excellent summaries are available, however (see Kautz, 1988, pp. 43-47; Davis and Kenyon, 1989, pp. 62-64; Suzuki and Knudtson, 1989, pp. 41-45).] In his book, *The Case Against Accident and Self-Organization,* Dean Overman provided the following excellent summary [see Figures 2 and 3 on the following pages].

> A DNA molecule is comprised of thousands of long chains of nucleotides (polynucleotides) each consisting of three parts. One part is the pentose or five carbon sugar known as deoxyribose. A second part is a phosphate group, and the third part is a nitrogen base of either adenine (A), guanine (G), cytosine (C) or thymine (T). Alternating sugar and phosphate molecules connect each nucleotide chain in a ladder type configuration coiled around a central axis in a twisted double spiral or helix. The two chains run in opposite directions with 10 nucleotides per turn of the helix. The rungs of the bases are pairs of either adenine and thymine (A-T) or cytosine with guanine (C-G). A relatively weak hydrogen bond connects these bases... (1997, p. 34).

Genes, then, are specific segments of DNA (although not all DNA assumes the form of genes; some resides in extranuclear organelles such as plasmids, and some is non-coding). Chromosomes—which consist of DNA and other material—are macromolecules composed of repeating nucleotides that serve as carriers for genes, with thousands of genes being aligned along each chromosome. [Not all human genes, however, are found on chromosomes; a few reside within mitochondria located in the cytoplasm; see Ridley, 1999, p. 9.] Each chromosome consists of a pair of long (roughly three feet), tightly coiled, double-stranded DNA molecules, with each chromosome possessing one long arm and one short arm separated by a middle "pinch point" known as a centromere.

Every living organism has a specified number of chromosomes in each of its somatic cells. A corn cell has 20; a mouse, 40; a gibbon, 44; and a human, 46. Germ cells in humans, however, have only 23 chromosomes each so that during the union of the male and female gametes, the total will be the standard human number of 46 (23 + 23). [Of these, 22 pairs are numbered in approximate order of size from the largest (#1) to the smallest (#22), while the remaining pair consists of the sex chromosomes: two large X chromosomes in women, one X and one small Y in men.] As a result, genes are in-

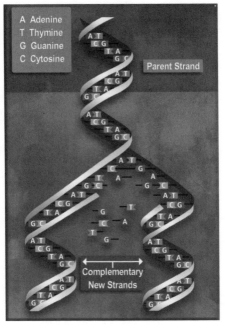

Figure 2 – DNA shown in double-helix, parent-strand form (top), and during replication of two new complementary strands (bottom). Source: U.S. Department of Energy Human Genome Program [on-line], http://www.ornl.gov/hgmis.

herited in pairs consisting of one portion from the father and one from the mother, thereby ensuring genetic diversity.

An average gene consists of about 1,000 nucleotides [Figure 3] that normally appear in triplets such as AGC or ATG (see Perloff, 1999, p. 72). While most triplets specify amino acid production, some function as a "stop" command, just as a telegram might contain "stop" to end a sentence. All living organisms—humans, animals, and plants—depend on this code for their existence. Furthermore, each gene is the blueprint the cell uses to assemble a protein that is composed of a long necklace of amino acids (with each protein consisting of a distinct sequence of those amino acids). [A typical protein contains approximately 300 amino acids (see Macer, 1990, p. 2).]

Thanks to the progress that has been made in both genetics and molecular biology, we now possess techniques by which it is possible to determine the exact chemical sequence of any gene from any organism. The **genotype** is the complete set of genes that the organism possesses—something determined at the time of conception for multicellular organisms. It is the same in all cells of an individual organism. The genotype of all cells derived from a particular cell will be the same, unless a mutation occurs. [It is estimated that 90% of all known gene

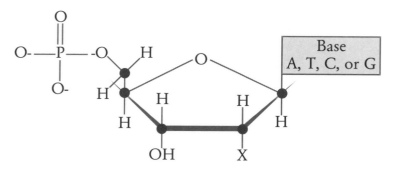

Figure 3 – The structure of a nucleotide. Circles represent carbon atoms. In DNA the sugar is deoxyribose, with a hydrogen atom at position X; in RNA the sugar is ribose, with a hydroxyl (OH) group at position X. In DNA, the base can be A,G,C, or T; in RNA, the base can be A,G,C, or U.

mutations occur in autosomal chromosomes (as opposed to sex chromosomes–see Macer, 1990, p. 4).] For organisms that reproduce sexually, the **genotype** of each new individual will be different since the genes from the two parents are combined. The **phenotype** of an individual is determined by the constant interaction of their genotype and the environment.

The DNA molecule truly is amazing, but it still has certain built-in limits. As geneticist Richard Lewontin remarked: "DNA is a dead molecule, among the most nonreactive, chemically inert molecules in the living world" (2000, p. 141). Matt Ridley referred to DNA as "a helpless, passive piece of mathematics, which catalyses no chemical reactions" (1999, p. 17). What is the point of such statements? Jonathan Wells has explained:

> Although molecular biology has demonstrated conclusively that DNA carries the genetic code for the amino acid sequences of proteins, this is not sufficient to specify a whole organism. Combining DNA with all the ingredients necessary for protein synthesis does not make a cell.... Molecular biology has shown that an organism's DNA specifies the building materials. It turns out, however, that **the assembly instructions are largely in other components of the cell**, and that the floor plan has not yet been discovered. So there are clearly other factors involved in heredity and development besides DNA (1998, pp. 62,64).

[This information will become important in separating fact from fiction in the discussion below on the Human Genome Project.]

Strictly speaking, of course, DNA is not actually a **self**-replicating molecule. As Lewontin explained:

> DNA has no power to reproduce itself. Rather it is produced out of elementary materials by a complex cellular machinery of proteins.... The newly manufactured DNA is certainly a **copy** of the old, and the dual structure of the DNA molecule provides a complementary template on which the copying process works...[but] no living molecule is self-reproducing (2000, p. 142, emp. in orig.).

DNA **does** replicate, however. And the process by which it does so is an enormously complex one with many different components that interact to ensure the faithful transfer of genetic information to the next generation. Biochemist Michael Behe noted:

> A large number of parts have to work together to that end. In the absence of one or more of a number of the components, DNA replication is either halted completely or significantly compromised, and the cell either dies or becomes quite sick (1998, p. 185).

What, then, is involved in reproducing the DNA molecule so that it can be passed from cell to cell and generation to generation?

Once the structure of DNA finally was elucidated, scientists discovered how, during cell division, the DNA is replicated to produce a genome [the organism's total genetic content] for each new daughter cell. The secret lies in the pairing of the bases—A to T, and G to C. During the replication process, the two complementary strands of DNA "unzip" down the middle. A new strand then begins to form alongside each of the originals, laying in an A wherever there is an opposing T, a T where there is an A, a G to a C, and a C to a G. The end result is two new double-stranded portions of DNA that, in most instances, are identical to the originals in their base sequences [see Figure 2]. Ridley described the process by comparing the genetic material to a book.

> The genome is a very clever book, because in the right conditions it can both photocopy itself and read itself. The photocopying is known as **replication**, and the reading as **translation**. Replication works because of an ingenious property of the four bases: A likes to pair with T, and G with C. So a single strand of DNA can copy itself by assembling a complementary strand with Ts opposite all the As, As opposite all the Ts, Cs opposite all the Gs and Gs opposite all the Cs. In fact, the usual state of DNA is the famous **double helix** of the original strand and its complementary pair intertwined.

To make a copy of the complementary strand therefore brings back the original text. So the sequence ACGT becomes TGCA in the copy, which transcribes back to ACGT in the copy of the copy. This enables DNA to replicate indefinitely, yet still contain the same information.

Translation is a little more complicated. First the text of a gene is **transcribed** into a copy by the same base-pairing process, but this time the copy is made not of DNA but of RNA, a very slightly different chemical.... This RNA copy, called the **messenger RNA**, is then edited....

The messenger is then befriended by a microscopic machine called a **ribosome**, itself made partly of RNA. The ribosome moves along the messenger, translating each three-letter codon in turn into one letter of a different alphabet, an alphabet of twenty different **amino acids**, each brought by a different version of a molecule called **transfer RNA**. Each amino acid is attached to the last to form a chain in the same order as the codons. When the whole message has been translated, the chain of amino acids folds itself up into a distinctive shape that depends on its sequence. It is now known as a **protein**.

Almost everything in the body, from hair to hormones, is either made of proteins or made by them. Every protein is a translated gene (1999, pp. 6,7,8, emp. in orig.).

Yes, the process described above is utterly amazing. But no less amazing is the fact that it takes place in a DNA fiber that is only two millionths of a millimeter thick (barely visible under an electron microscope). Yet the amount of information contained within it "is so immense in the case of human DNA that it would stretch from the North Pole to the equator if it was typed on paper, using standard letter sizes" (Gitt, 1997, p. 90). As Anderson observed: "If the tightly coiled DNA strands inside a single human adult were unwound and stretched out straight, they would cover the distance to the moon half a million times. Yet when coiled, all the strands could fit inside a teaspoon" (1980, p. 50).

The DNA molecule must be incredibly stable, since the genetic information stored within it may need to function in a living organism for up to a century or more. It also must be completely reproducible so that its complex informational content can be passed successfully from generation to generation. As it turns out, DNA does, in fact, possess each of these traits, and thereby fulfills the necessary and essential criteria of stability and replicability. Are we to be convinced, however, that all of this occurred merely **by chance**?

ORIGIN OF THE GENETIC CODE

The nucleic acid-based genetic code—with its complexity, orderliness, and function—provides the most powerful kind of evidence for **intelligent design**, which requires a Designer. But whence has it come?

Since the elucidation of the genetic code in the mid-1950s, materialists have suggested that those mythical parents, "father time" and "mother nature," gave birth to the genetic code via purely chance processes. As Nobel laureate Jacques Monod put it: "Chance alone is the source of every innovation, of all creation in the biosphere.... All forms of life are the product of chance..." (1972, pp. 110,167). Such a view, however, ascribes to "chance" properties that it does not, and cannot, possess. Sproul, Gerstner, and Lindsley addressed this logical fallacy and concluded:

> Chance is incapable of creating a single molecule, let alone an entire universe. Why not? Chance is no thing. It is not an entity. It has no being, no power, no force. It can effect nothing for it has no causal power within it (1984, p. 118).

Chance cannot create. And it certainly cannot create something as complex as the genetic code. Furthermore, as science writer Matt Ridley observed: "DNA is information, a message written in a code of chemicals" (1999, p. 13). And, as information scientist Werner Gitt correctly noted: "Coding systems are not created arbitrarily, but they are optimized according to criteria.... Devising a code is a **creative mental**

process. Matter can be a **carrier** of codes, but it cannot **generate** codes" (1997, pp. 59,67, emp. added). Whence, then, has come the genetic code? What "creative mental process" imposed the information on it that it contains? In their textbook, *The New Biology,* evolutionists Robert Augros and George Stanciu wrote:

> What cause is responsible for the origin of the genetic code and directs it to produce animal and plant species? It cannot be matter because of itself matter has no inclination to these forms.... **There must be a cause apart from matter** that is able to shape and direct matter. Is there anything in our experience like this? Yes, there is: our own minds. The statue's form originates in the mind of the artist, who then subsequently shapes matter, in the appropriate way.... **For the same reasons there must be a mind that directs and shapes matter in organic forms** (1987, p. 191, emp. added).

In speaking of the origin of the genetic code, and the simultaneous appearance of the decoding mechanism that accompanies it, evolutionist Caryl Haskins lamented: "By a pre-Darwinian (or a skeptic of evolution after Darwin) **this puzzle would surely have been interpreted as the most powerful sort of evidence for special creation**" (1971, 59:305, emp. added, parenthetical comment in orig.). The late evolutionist Carl Sagan of Cornell University admitted:

> The number of possible ways of putting nucleotides together in a chromosome is enormous. Thus **a human being is an extraordinarily improbable object**. Most of the $10^{2.4 \times 10^9}$ possible sequences of nucleotides would lead to complete biological malfunction (1997, 22:967, emp. added).

Sir Francis Crick therefore observed:

> An honest man, armed with all the knowledge available to us now, could only state that in some sense, **the origin of life appears at the moment to be almost a miracle**, so many are the conditions which would have had to have been satisfied to get it going (1981, p. 88, emp. added).

Wilder-Smith offered the following observation about the origin of the genetic code.

> The almost unimaginable complexity of the information on the genetic code along with the simplicity of its concept (four letters made of simple chemical molecules), together with its extreme compactness, **imply an inconceivably high intelligence behind it**. Present-day information theory permits no other interpretation of the facts of the genetic code (1976, pp. 258-259, emp. added).

This is the very point that Gitt made in his 1997 book on information theory when he wrote: "The coding system used for living beings is optimal from an engineering standpoint. This fact strengthens the argument that it was a case of **purposeful design** rather than fortuitous chance" (p. 95, emp. added). British evolutionist Richard Dawkins once observed: "The more statistically improbable a thing is, the less we can believe that it just happened by blind chance. Superficially the obvious alternative to chance is an intelligent Designer" (1982, p. 130). I suggest, however, that since the genetic code "appears to be almost a miracle" which "implies an inconceivably high intelligence behind it," then it hardly is "superficial" to believe that it must have had a designer.

FUNCTION AND DESIGN OF THE GENETIC CODE

Faithful, accurate cellular division is critically important, of course, because without it life could not continue. In his presidential address to the British Association for the Advancement of Science, William Bateson made this startling admission: "Descent used to be described in terms of blood. Truer notions of genetic physiology are given by the Hebrew expression 'seed.' If we say he is 'of the seed of Abraham,' we feel something of the **permanence and indestructibility** of that germ which can be divided and scattered among nations, but remains recognizable in type and characteristic after 4,000 years" (1914, emp. in orig.). Seventy-five years later, not much had changed. Suzuki and Knudtson commented, for example:

Yet long before the concept of the "gene" crystallized in human consciousness early in this century, human beings felt compelled to search for ways to make sense of at least the most visible evidence of biological inheritance that surrounded them. For they could not help noticing the recurring pattern of reproduction in the natural world by which every form of life seemed to generate new life–"according to its own kind." The keen-eyed agriculturalists among them could not have missed the similarity between successive generations of livestock and crops. Nor was it possible to ignore the sometimes uncanny resemblances between members of one's own immediate family or ancestral lineage (1989, p. 32).

Suzuki and Knudtson went on to suggest, however, that these poor humans lived in a state of "scientific innocence," and thus could be excused for not knowing any better. But is it necessarily a state of "scientific innocence" to rely on empirical observations and common sense? John Gribbin, himself an evolutionist, has admitted that "...once a fertilized, single human cell begins to develop, the original plans are **faithfully copied** each time the cell divides (a process called mitosis) so that every one of the thousand million million cells in my body, and in yours, contains a **perfect replica** of the original plans for the whole body" (1981, p. 193, emp. added, parenthetical comment in orig.). Wilder-Smith noted:

The Nobel laureate, F.H. Crick has said that if one were to translate the coded information on one human cell into book form, one would require one thousand volumes each of five hundred pages to do so. And yet the mechanism of a cell can copy faithfully at cell division all this information of one thousand volumes each of five hundred pages in just twenty minutes (1976, p. 258, emp. added).

Sparrows produce nothing but sparrows and human beings produce nothing but human beings because all organisms faithfully reproduce copies of their own genetic code. Dr. Bateson spoke of the **permanence and indestructibility** of the "seed." Dr. Gribbin said the code is copied **faithfully**.

Suzuki and Knudtson commented on the **recurring pattern of reproduction**. It matters little what terms these evolutionists use; their point is still clear—all living things reproduce "after their kind."

However, while it is important to recognize that although "faithful reproduction" at the cellular level is essential, life could not sustain itself without the existence and continuation of the extremely intricate genetic code contained within each cell. Scientific studies have shown that the hereditary information contained in the code found within the nucleus of the living cell is universal in nature. Regardless of their respective views on origins, all scientists acknowledge this. Evolutionist Richard Dawkins observed: "The genetic code is universal.... The complete word-for-word universality of the genetic dictionary is, for the taxonomist, too much of a good thing" (1986, p. 270). Creationist Darrel Kautz agreed: "It is recognized by molecular biologists that the genetic code is universal, irrespective of how different living things are in their external appearances" (1988, p. 44). Or, as Matt Ridley put it in his 1999 book, *Genome:*

> Wherever you go in the world, whatever animal, plant, bug or blob you look at, if it is alive, it will use the same dictionary and know the same code. **All life is one**. The genetic code, barring a few tiny local aberrations, mostly for unexplained reasons in the ciliate protozoa, is the same in every creature. We all use exactly the same language.
>
> **This means—and religious people might find this a useful argument—that there was only one creation, one single event when life was born**.... The unity of life is an empirical fact (pp. 21-22, emp. added).

It is the genetic code which ensures that living things reproduce faithfully "after their kind," exactly as the principles of genetics state that they should. Such faithful reproduction, of course, is due both to the immense complexity and the intricate design of that code. It is doubtful that anyone cognizant of the facts would speak of the "simple" genetic code. A.G. Cairns-Smith has explained why:

Every organism has in it a store of what is called **genetic information**.... I will refer to an organism's genetic information store as its **Library**.... Where is the Library in such a multicellular organism? The answer is everywhere. With a few exceptions every cell in a multicellular organism has a complete set of all the books in the Library. As such an organism grows, its cells multiply and in the process the complete central Library gets copied again and again.... The human Library has 46 of these cord-like books in it. They are called chromosomes. They are not all of the same size, but an average one has the equivalent of about 20,000 pages.... Man's Library, for example, consists of a set of construction and service manuals that run to the equivalent of about a million book-pages together (1985, pp. 9,10, emp. in orig.).

Wilder-Smith concurred with such an assessment when he wrote:

> Now, when we are confronted with the genetic code, we are astounded at once at its simplicity, complexity and the mass of information contained in it. One cannot avoid being awed at the sheer density of information contained in such a miniaturized space. When one considers that the entire chemical information required to construct a man, elephant, frog, or an orchid was compressed into two minuscule reproductive cells, one can only be astounded. **Only a sub-human could not be astounded**. The almost inconceivably complex information needed to synthesize a man, plant, or a crocodile from air, sunlight, organic substances, carbon dioxide and minerals is contained in these two tiny cells. If one were to request an engineer to accomplish this feat of information miniaturization, one would be considered fit for the psychiatric line (1976, pp. 257-259, emp. in orig.).

It is no less amazing to learn that even what some would call "simple" cells (e.g., bacteria) have extremely large and complex "libraries" of genetic information stored within them. For example, the bacterium *Escherichia coli*, which is by no means the "simplest" cell known, is a tiny rod only a thousandth of a millimeter across and about twice as long, yet "it is an indication of the sheer complexity of *E. coli* that its Library runs to a thousand page-equivalent" (Cairns-Smith, p. 11). Biochemist Michael Behe has suggested that the amount of DNA in a cell "varies roughly with the complexity of the organism" (1998, p. 185). There are notable exceptions, however. Humans, for example, have about 100 times more of the genetic-code-bearing molecule (DNA) than bacteria, yet salamanders, which are amphibians, have 20 times more DNA than humans (see Hitching, 1982, p. 75). Humans have roughly 30 times more DNA than some insects, yet less than half that of certain other insects (see Spetner, 1997, p. 28).

It does not take much convincing, beyond facts such as these, to see that the genetic code is characterized by orderliness, complexity, and adeptness in function. The order and complexity themselves are nothing short of phenomenal. But the **function** of this code is perhaps its most impressive feature, as Wilder-Smith explained when he suggested that the coded information

> ...may be compared to a book or to a video or audio-tape, with an extra factor coded into it enabling the genetic information, under certain environmental conditions, to read itself and then to execute the information it reads. It resembles, that is, a hypothetical architect's plan of a house, which plan not only contains the information on how to build the house, but which can, when thrown into the garden, build entirely of its own initiative the house all on its own without the need for contractors or any other outside building agents.... Thus, it is fair to say that the **technology** exhibited by the genetic code is orders of magnitude higher than any technology man has, until now, developed. What is its secret? The secret lies in its ability to store and to execute incredible magnitudes of conceptual information in the ultimate molecular miniaturization of the information storage and retrieval system of the nucleotides and their sequences (1987, p. 73, emp. in orig.).

This "ability to store and to execute incredible magnitudes of conceptual information" is where DNA comes into play.

Wilder-Smith concluded: "The information stored on the DNA molecule is that which controls totally, as far as we at present know, by its interaction with its environment, the development of all biological organisms" (1987, p. 73). E.H. Andrews summarized how this can be true:

> The way the DNA code works is this. The DNA molecule is like a template or pattern for the making of other molecules called "proteins...." These proteins then control the growth and activity of the cell which, in turn, controls the growth and activity of the whole organism (1978, p. 28).

Thus, the DNA contains the information that allows proteins to be manufactured, and the proteins control cell growth and function, which ultimately are responsible for each organism. The genetic code, as found within the DNA molecule, is vital to life as we know it. In his book, *Let Us Make Man*, Bruce Anderson referred to it as "the chief executive of the cell in which it resides, giving chemical commands to control everything that keeps the cell alive and functioning" (1980, p. 50). Kautz followed this same line of thinking when he stated:

> The information in DNA is sufficient for directing and controlling all the processes which transpire within a cell including diagnosing, repairing, and replicating the cell. Think of an architectural blueprint having the capacity of actually building the structure depicted on the blueprint, of maintaining that structure in good repair, and even replicating it (1988, p. 44).

Likely, many people have not considered the exact terminology with which the genetic code is described in the scientific literature. Lester and Bohlin observed:

> The DNA in living cells contains coded information. It is not surprising that so many of the terms used in describing DNA and its functions are language terms. We speak of the genetic **code**. DNA is **transcribed** into RNA. RNA is **translated** into protein.... Such designations are not simply convenient or just anthropomorphisms. They accurately describe the situation (1984, pp. 85-86, emp. in orig.).

How, then, did this complex chemical code arise? What "outside source" imposed the information on the DNA molecule?

IMPLICATIONS OF THE
HUMAN GENOME PROJECT

On Monday, June 26, 2000, the President of the United States and the Prime Minister of Great Britain jointly called a press conference that not only received instantaneous, worldwide news coverage, but also captured the attention of people around the globe (see Office of Technology Policy, 2000). As the ambassadors of Japan, Germany, and France watched (along with some of the planet's most distinguished scientists, who had joined them either in person or via satellite), the two world leaders announced what one science writer called "the greatest intellectual moment in history, bar none!"–the deciphering of the code contained in the entire human genome.

The news media–both popular and scientific–had a field day. The July 3, 2000 bright red cover of *Time* magazine screamed in huge, yellow letters–"Cracking the Code!" Upon opening the magazine to read the text of the cover story, the reader was met with an audacious headline in giant type that announced: "The Race Is Over!" The July 3 issue of *U.S. News & World Report* covered the story under the heading, "We've Only Just Begun" (Fischer, 2000, 129[1]:47). One week later, in its July 10 issue, *U.S. News & World Report* assigned its highly touted editor-at-large, David Gergen, to write an editorial that was titled "Collaboration? Very Cool" about the success of the joint effort (2000, 129[2]:64). The July 3 issue of *Newsweek* contained a feature article, "A Genome Milestone," discussing the project (Hayden, 2000, 129[1]:51). The June 30 issue of *Science*, the official organ of the American Association for the Advancement of Science (Marshall, 2000, 288:2294-2295), and the June 29 issue of *Nature,* the official organ of its counterpart in Great Britain, the British Association for the Advancement of Science (Macilwain, 2000, 405:983-984), each devoted in-depth stories to the "cracking of the code." The

July 2000 issue of *Scientific American* also weighed in (Brown, 2000, 283[1]:50-55), as did numerous other professional journals in countries on almost every continent.

Emotional exhilaration ran high, and descriptive adjectives flowed freely. Professional writers, as well as some of the scientists involved in the events that led to the decoding of the human genome, variously described the results as the "holy grail" of biology and "the most important scientific effort that mankind has ever mounted"—and did not hesitate to compare the saga to the Manhattan Project that developed the atomic bomb in the mid-1940s or the Apollo Project that landed men on the moon on July 20, 1969. *Time's* cover-story authors remarked authoritatively: "It's impossible to overstate the significance of this achievement" (Golden and Lemonick, 2000, 156[1]:19).

Amidst all the hoopla, important questions are bound to arise. For example, what, exactly, is the human genome? What do scientists mean when they say they have "decoded" it? What do these events mean for mankind—either now or in the future? What are the potential benefits and/or drawbacks associated with such research? When can humanity expect to experience them? Are there any scientific, ethical, or moral implications to be considered? If so, what are they and how should we handle them? And what are the implications of the project in the creation/evolution controversy? These kinds of questions often accompany the invention and development of major new scientific technologies, and deserve a well-reasoned, informed response.

Whenever the President of the United States and the Prime Minister of Great Britain call a news conference that is broadcast worldwide in order to discuss a **scientific** matter, it must be pretty heady stuff. What, exactly, is the Human Genome Project? Why has it generated such tremendous publicity of late? Is all the hoopla surrounding it justified—or even correct?

An organism's genome is its total genetic content. [The phrase "nuclear genome" refers solely to the DNA within the nucleus; the phrase "human genome" refers to all of the DNA contained in an entire human (haploid) cell, rather than just that in the nucleus.] In the late 1980s, scientists began discussing the possibility of obtaining a detailed map and complete DNA sequence of the genome of a variety of organisms, including the bacterium *Escherichia coli*, the yeast *Saccharomyces cerevisiae*, the roundworm *Caenorhabditis elegans*, the fruit fly *Drosophila melanogaster* (all of which, by the way, had been completed by the end of 1999), the mouse, wheat, rice, and of course, *Homo sapiens*. [For an update on the progress regarding the sequencing of the genome of the mouse and other species, see Karow, 2000, 283[1]:53.]

The mere thought of mapping **all** the chromosomes and sequencing **all** the genes of even a "simple" living organism should be enough to send chills down the spine of every hardworking molecular biologist. After all, a bacterium can have **4 million** nucleotide bases in its genetic repertoire, while more complicated organisms such as human beings can possess more than **3 billion**. And, curiously, some amphibians and flowering plants have more than 10 times the number of nucleotide bases found in human beings (see Roth, 1998, p. 70; Avers, 1989, pp. 142-143; Fraser, et al., 1995, 270:397-403; Goffeau, 1995, 270:445-446). But, by the beginning of the year 2000, the genome sequences of more than 20 species had been published on the Internet, and the one-billionth base of human DNA had been sequenced (see Macer, 2000). Erika Check, writing in the August 14, 2000 issue of *Newsweek*, quoted Claire Fraser, head of the Institute for Genomic Research, who suggested that within the next year or so scientists will begin decoding the genomes of the top twenty human pathogens [disease-causing organisms] (136[7]:9). [In fact, in its July 13, 2000 issue, *Nature* reported that scientists in the country of Brazil had just completed the "first sequence of a free-living plant pathogen" and that their paper (published in that week's issue of the journal) represented "a significant scientific milestone" (see Editorial, 2000a, 406:109; see also Simpson, et al., 2000, 406:151-156). Less than three weeks later, *Nature* announced in its August 3, 2000 issue that the genes of *Vibrio cholerae*, the microorganism that causes cholera, had been completely sequenced (see: Heidelberg, et al., 2000, 406:477-483; Check, 2000, 136[7]:9).]

In 1990, the Human Genome Project [or HGP; also sometimes referred to as the Human Genome Initiative] began (see Collins, 1997, p. 98). The name is a collective moniker for several projects that actually began in the late 1980s in several countries, following a decision by the United States Department of Energy [DOE] to: (a) create an ordered set of DNA segments from known chromosomal locations; (b) develop new computational methods for analyzing genetic map and DNA sequence data; and (3) develop new instruments and techniques for detecting and analyzing DNA (see Office of Technology Assessment, 1988). However, some in the biological community were a bit wary of DOE physicists "doing biology." Thus, because the National Institutes of Health [NIH] is the major funder of biomedical research in America, its scientists signed on to join the project. [Francis Collins, M.D., Ph.D., is the head of the U.S. Human Genome Project.]

Shortly after the formation of the HGP in the United States, scientists from several foreign countries were invited to join in the effort, which resulted in the formation of the HGP international analogue–the Human Genome Organization [HUGO]. Included in the international effort were scientists from France, Great Britain, Japan, and elsewhere. In 1991 the Human Genome Diversity Project [HGDP] was begun, with a mandate to collect DNA samples for analysis from at least 25 unrelated individuals in 400 different populations around the world. Dr. Luigi Cavalli-Sforza, professor emeritus of genetics at Stanford University, heads the program (see Macer, 2000; Cavalli-Sforza, 2000, p. 69). In mid-1999, British science writer Matt Ridley wrote in his book, *Genome:*

> Being able to read the genome will tell us more about... our nature and our minds than all the efforts of science to date. It will revolutionise anthropology, psychology, medicine, palaeontology and virtually every other science.... **Some time in the year 2000, we shall probably have a rough first draft of the**

complete human genome. In just a few short years we will have moved from knowing almost nothing about our genes to knowing everything. I genuinely believe that we are living through the greatest intellectual moment in history. Bar none (p. 5, emp. added).

Ridley's prediction has come true. The HGP now has achieved one of its main goals—producing a "rough first draft" of the human genome. Two groups—one governmental [the HGP] and one from corporate America [Celera Genomics, headed by its CEO, Dr. Craig Venter]—had been pursuing the goal of mapping the entire human genome independently of each other. [On January 10, 2000, for example, scientists at Celera announced they had sequences equal to over 90% of the human genome, and 97% of all genes, in their database (see Editorial, 2000b).] Eventually, however, the two groups agreed to work together. And work they did! On June 26, 2000, the announcement was made that, for all practical purposes, the mapping of the human genome was complete. In its cover story the following week (July 3), *Time* magazine reported on the meaning and importance of the announcement.

> After more than a decade of dreaming, planning and heroic number crunching, both groups have deciphered essentially all the 3.1 billion biochemical "letters" of human DNA, the coded instructions for building and operating a fully functional human....

Armed with the genetic code, scientists can now start teasing out the secrets of human health and disease at the molecular level—secrets that will lead at the very least to a revolution in diagnosing and treating everything from Alzheimer's to heart disease to cancer, and more (Golden and Lemonick, 2000, 156[1]:19-20).

The Human Genome Project is set up to proceed in two distinct stages, the first of which is that of "physical mapping." This phase will examine short stretches of DNA in order to determine sequences along each chromosome as "landmarks" (somewhat like the mile markers found along U.S. interstate highways). These markers then will be of importance in finding exactly where, along each chromosome, particular genes reside. In the second phase of the project, various laboratories will examine an entire chromosome (or section of a chromosome, depending on its size) in order to determine the complete ordered sequence of nucleotides in its DNA. It is after this critical second phase, to use the words of Harvard's Lewontin, "that the fun begins, for biological sense will have to be made, if possible, of the mind-numbing sequence of three billion A's, T's, C's, and G's" (2000, p. 162).

Truth be told, the processing of making "biological sense" out of the human genome already has begun in earnest. The December 2, 1999 issue of *Nature* announced, for example, that the first human chromosome (#22) had been completely sequenced (see Little, 1999, 402:467-468; Dunham, et al., 1999, 402:489-495; Donn, 1999). And in May 2000, the HGP announced that it not only had completed its own working draft of chromosome 22, but also had completed the sequencing of chromosome 21, which is involved with Down's syndrome and several other diseases (see Brown, 2000, 283[1]:50-55; for a full account of the chromosome 21 story, see *Scientific American's* Web site at http://www.sciam.com/explorations/200 0051500chrom21).

But where, exactly, is the HGP now? Almost all of the genome data already are being used. As of June 2000, 85% of the human genome was available on the World Wide Web (see Regalado, 2000, 103[4]:97-98). On February 16, 2001, a special issue of *Science* was devoted almost entirely to the human genome. In that report, scientists revealed that the human genome consisted of 2.91 billion nucleotide base pairs. However, this rough draft was accomplished using a "shotgun" approach to the entire genome, and as such, there were many gaps left to fill. Since that time, researchers have been slogging away to collect data from those areas not examined by the initial survey.

On April 14, 2003, the International Human Genome Consortium announced the successful completion of the Human Genome Project—more than two years ahead of schedule. The press report read as follows: **"The human genome is com-**

plete and the Human Genome Project is over" (see "Human Genome Project...," 2003, emp. added). This particular announcement came almost fifty years to the day after James Watson and Francis Crick unveiled their description of the DNA double helix.

As Dr. Francis Collins, director of the genome center at the National Institutes of Health, noted: "The completion of the Human Genome Project should not be viewed as an end in itself. Rather, it marks the start of an exciting new era—the era of the genome in medicine and health" (as quoted in "Human Genome Project...," 2003). The emphasis now will be placed on how we can use the information we have obtained thus far.

While the April 14, 2003 announcement does indeed mark a milestone in science, it has not come without some criticism. The finished sequence produced by the Human Genome Project covers only about 98-99% of the human genome's gene-containing regions—thus the word "complete" may be somewhat premature. The working draft that was reported in June 2000 covered only 90 of the gene-containing regions; thus, this finished product is considerably more complete (and more accurate) than the draft version. The remaining gaps represent the regions of DNA that scientists have found difficult to sequence reliably. Elbert Branscom noted: "It's the best effort that mortals can do with current technology" (as quoted in Pearson, 2003). Reporting on these missing gaps, Nicholas Wade noted:

> When the working draft of the human genome was produced, consortium scientists called it the "Book of Life," with each chromosome a chapter. In the edition published today, small sections at the beginning, end, and middle of each chapter are blank, along with some 400 assorted paragraphs whose text is missing, although the length of the missing passages is known (2003).

Nicholas went on to quote Evan Eichler, a computational biologist at Case Western Reserve University who studies certain duplicated regions of the genome. Dr. Eichler observed that this was indeed a "momentous achievement," but that "we shouldn't declare a job 'complete' until it is." He went on to note that it was "critical that the complete human genome sequence be, well, complete, in the fullness of time" (as quoted in Wade, 2003). Some scientists speculate that it could take an additional 10-20 years to sequence the unusual structures that remain unknown in the human genome. In addition, researchers now are anxious to compare the human genome to that of animals, so the race is on to "finish the job" and complete the genomes of a variety of animals.

With this first major step out of the way, biologists now must systematically identify the regions of DNA that hold genes of interest. In the April 24, 2003 issue of *Nature*, the National Human Genome Research Institute (NGHRI) officially unveiled its vision for the future of genomic research, thereby officially closing one door and opening a new one. There is much we still do not know, and much work yet to be done.

Eric Lander, who is the director of the Whitehead Institute for Biomedical Research/MIT center for Genome Research (the world's most productive academic gene sequencing facility and the flagship of the international Human Genome Project), admitted in an interview:

> The truth is that the human genome is going to have all kinds of nasty little bits that are hard to fill in at the end: the middles of chromosomes, called the centromeres, the ends of chromosomes, called the telomeres, and so on. This is not like the transcontinental railroad, where at some point someone is going to nail the golden spike, and then and only then can you go cross-country. There is no golden nucleotide to be nailed into the double helix at the end....

> The genome is a very elaborate program, and we don't know how to read it. It's as if we have some ancient computer code that was written...years ago and now we are trying to figure out what it does. I think what biologists are going to be doing for the next decade is figuring out the circuitry of the genome by monitoring how the 50,000 to 100,000 genes are turned on and off and how all the proteins come on and off in the cell (as quoted in Regalado, 2000, 103[4]:97-98).

In other words, while we now know what each of the letters **is**, we still have to determine what each letter **does**–i.e., what each one is responsible for. Specific segments of DNA make up genes that control specific things (such as hair color, eye color, etc.). What each of those genes controls is still a mystery. A good analogy might be a young child who has just learned the alphabet. While that is a great accomplishment, that child has much to learn before he or she can read and understand a novel. Scientists are now in the position of that young child. We now know the "alphabet" of the human genome. And we can even "read" many words (genes). But we are a long way from reading and comprehending the entire book. Thus, scientists are now faced with the Herculean task of finding out what each gene does, inventing tools that can inexpensively screen the entire genome of humans, and then discovering tools that will allow them to alter those genomes that contain things such as genes contributing to conditions like diabetes, heart disease, or mental illness.

"ERROR MESSAGES"– SNPS AND MUTATIONS

As much as we might wish it were true, mapping the DNA sequence of a single human–or even many humans from populations around the world–will not produce an accurate map of **a** human genome. Why not? The reason has to do with what geneticists refer to as "single nucleotide polymorphisms" (known as SNPS–pronounced SNIPS). Although human DNA is "almost" the same from every person on Earth, it is not **exactly** the same. The fact is, there is an approximate 0.1% variation in the nucleotides that compose human DNA. Generally, such variation is caused by a single nucleotide–thus the name "single nucleotide" polymorphisms [poly–many; morphisms–forms]. The DNA being sequenced in the HGP actually is a composite of human tissue cell lines from several people. As Lemonick wrote in his *Time* article:

Scientists...are putting together databases of tissue samples to look for one-letter genetic differences.... Both the Human Genome Project and Celera are currently sequencing the genomes of many different people, of both sexes and all sorts of ethnic backgrounds, to get a better sense of where the SNPS are (2000, 156[1]:28).

But, as *Newsweek's* Thomas Hayden has reminded us: "Meanwhile, the benefits of genomic research—from predicting risk for hereditary disease to developing new drugs designed for an individual's genetic makeup—are still years away..." (2000, 136[1]:51). One scientist, Richard K. Wilson of Washington University (a partner in the public consortium of the Human Genome Project), plainly admitted in an interview in the July 2000 issue of *Scientific American*:

> For a long time, there was a big misconception that when the DNA sequencing was done, we'd have total enlightenment about who we are, why we get sick and why we get old. Well, total enlightenment is decades away (as quoted in Brown, 2000, 283[1]:50).

Luigi Cavalli-Sforza, director of the Human Genome Diversity Project that is examining DNA samples from over 400 populations worldwide, has explained why accurate knowledge of SNPS is so critical.

> If we take the DNA from one sperm (or egg) and compare it to the DNA of another random one, we find that there is on average one different nucleotide pair every thousand nucleotide pairs. There are therefore at least three million differences between the DNA in one sperm or egg and the DNA in another. All these differences originated by mutation, a spontaneous error made while copying DNA, which most frequently involves the replacement of one nucleotide by another of the four.... New mutations are therefore transmitted from parents to children.... A change in DNA may cause a change in a protein... (2000, pp. 68,17, emp. added).

And a change in a protein within a living system can herald severe problems. Organisms contain thousands of proteins that most often are composed of 300 or more amino acids linked together in chain-like fashion. Substitution of even a single amino acid at a critical position can be lethal (see Roth, 1998, p. 69; Radman, 1988, 259[2]:40-46). In an article in *Nature* titled "The Book of Genes," Peter Little explained why SNPS are so important within the context of the Human Genome Project.

> There is a general consensus that SNPs are probably the cause of most common genetic disorders. We all carry many SNPS but if we are unlucky enough to carry the "wrong" set of changes, we are predisposed to one or other of the common disorders with a genetic component such as diabetes, heart disease, asthma, or cancers.... If knowledge of gene differences can be combined with an understanding of the richness of environmental influences, we will have the key to unlocking the cause of most of the common disorders that kill or otherwise cause suffering (1999, 402:467-468).

In our day and age, of course, "neo-Darwinism" and the "modern synthetic theory" of Darwinism are in vogue. Neo-Darwinism, as its name implies, has added something "new" to the old theory of Darwinian evolution that was supposed to have occurred solely by natural selection. The "new" is genetic mutations. As Simpson and his co-authors suggested over four decades ago: "Mutations are the ultimate raw materials for evolution" (1957, p. 430). Forty-three years later, the view was still the same. In his 2000 book, *Quantum Evolution,* John J. McFadden wrote:

> Over millions of years, organisms will evolve by selection of mutant offspring which are fitter than their parents. Mutations are therefore the elusive source of the variation that Darwin needed to complete his theory of evolution. They provide the raw material for all evolutionary change (p. 65).

Currently, it is thought that evolution proceeds through the combined efforts of natural selection and genetic mutations. As Dr. McFadden went on to note: "Natural selection tends to favour organisms carrying advantageous mutations that allow them to produce more offspring" (p. 65). However, the true facts of science tell a story not in accord with such explanations—or with the concept of evolution they are intended to bolster. The whole point is this: mutations—yes; evolution—no. Consider why.

First, natural selection ("survival of the fittest") is a tautologous concept (i.e., employs circular reasoning). It simply requires the "fittest" organisms to leave the most offspring, and then at the same time defines the "fittest" organisms as those that leave the most offspring. Arthur Koestler, a vitalist philosopher, described the tautology of natural selection as follows:

> Once upon a time, it all looked so simple. Nature rewarded the fit with the carrot of survival and punished the unfit with the stick of extinction. The trouble only started when it came to defining fitness.... Thus natural selection looks after the survival and reproduction of the fittest, and the fittest are those which have the highest rate of reproduction.... [W]e are caught in a circular argument which completely begs the question of what makes evolution evolve (1978, p. 170).

Norman Macbeth, the Harvard-trained lawyer who authored the classic text, *Darwin Retried*, agreed with Koestler's assessment. "In the meantime, the educated public continues to believe that Darwin has provided all the relevant answers by the magic formula of random mutations plus natural selection—quite unaware of the fact that random mutations turned out to be irrelevant and natural selection a tautology" (1982, 2:18).

G.A. Peseley, in an article on "The Epistemological Status of Natural Selection," wrote:

> Evolution depends upon natural selection. Yet natural selection ("survival of the fittest") is a tautology (i.e., uses circular reasoning). It simply requires the "fittest" organisms to leave the most offspring, and then identifies the "fittest" organisms as those that leave the most offspring.

> One of the most frequent objections against the theory of natural selection is that it is a sophisticated tautology. Most evolutionary biologists seem uncon-

cerned about the charge and make only a token effort to explain the tautology away. The remainder, such as Professors Waddington and Simpson, will simply concede the fact. For them, natural selection is a tautology which states a heretofore unrecognized relation: The fittest—defined as those who will leave the most offspring—will leave the most offspring.

What is most unsettling is that some evolutionary biologists have no qualms about proposing tautologies as explanations. One would immediately reject any lexicographer who tried to define a word by the same word, or a thinker who merely restated his proposition, or any other instance of gross redundancy; yet no one seems scandalized that men of science should be satisfied with a major principle which is no more than a tautology (1982, 38:74).

The eminent Swedish botanist, Søren Løvtrup, in his book, *Darwinism: The Refutation of a Myth*, was even more harsh in his assessment.

After this step-wise elimination, only one possibility remains: **the Darwinian theory of natural selection, whether or not coupled with Mendelism, is false.** I have already shown that the arguments advanced by the early champions were not very compelling, and that there are now considerable numbers of empirical facts which do not fit with the theory. Hence, **to all intents and purposes the theory has been falsified**, so why has it not been abandoned? I think the answer to this question is that current evolutionists follow Darwin's example—they refuse to accept falsifying evidence (1987, p. 352, emp. added).

Or, as Colin Patterson of the British Museum of Natural History put it:

No one has ever produced a species by mechanisms of natural selection. No one has ever gotten near it and most of the current argument in neo-Darwinism is about this question: how a species originates. And it is there that natural selection seems to be fading out, and chance mechanisms of one sort of another are being invoked (1982).

Thus, natural selection does not provide a testable explanation of how mutations could produce more fit organisms (see Popper, 1975, p. 242).

Second, even evolutionists themselves admit that mutations, just like SNPS, are "errors" in DNA replication (see Ayala, 1978, 239 [3]:56-69). And these "errors," which (to quote Dr. Cavalli-Sforza) "can herald severe problems"—almost always are harmful. We know today, of course, that there are at least three **possible** kinds of mutations: (1) bad; (2) good; and (3) neutral. Neutral mutations are of no value, as they have, in essence, no "net effect." What, then, may be said about the **bad** or **good** mutations? Of the remainder of all mutations (after neutral ones have been eliminated), 99% of all remaining mutations are harmful (Winchester, 1951, p. 228; Martin, 1953, p. 100; Ayala, 1968, p. 1436; Morris, 1984, p. 203; Klotz, 1985, p. 181). As long ago as 1937, famed evolutionist Ernest Hooton observed:

> Saltatory evolution by way of mutation is a very convenient way of bridging over gaps between animal forms.... Now I am afraid that many anthropologists (including myself) have sinned against genetic science and are leaning upon a broken reed when we depend upon mutations (1937, p. 118).

Why did Dr. Hooton offer such an assessment? Mutations are known to be random, and under most conditions are destructive or even lethal to the individual in which they are expressed. After all, mutations are changes ("errors") in the DNA. As one evolutionist stated: "As a degenerative principle, providing the raw material for natural selection, random mutation is inadequate both in scope and theoretical grounding" (Wicken, 1979, p. 349). In other words, mutations, being random, cannot "order" anything, or make anything more complex. Natural selection can serve only to "weed out" those mutations that are harmful, at best preserving the "status quo." Or, as Koestler put it:

> In the meantime, the educated public continues to believe that Darwin has provided all the relevant answers by the magic formula of random mutations plus

natural selection—quite unaware of the fact that random mutations turned out to be irrelevant and natural selection a tautology (1978, p. 170).

Pierre-Paul Grassé, whom I already have quoted, is not a creationist and is, in fact, France's leading zoologist, having held the Chair of Evolution at the Sorbonne for twenty years. His opinion of mutations, as an explanatory mechanism of evolution, is this:

> Some contemporary biologists, as soon as they observe a mutation, talk about evolution. They are implicitly supporting the following syllogism (argument): mutations are the only evolutionary variations, all living beings undergo mutations, therefore all living beings evolve. This logical scheme is, however unacceptable: first, because its major premise is neither obvious nor general; second, because its conclusion does not agree with the facts. No matter how numerous they may be, mutations do not produce any kind of evolution.... The opportune appearance of mutations permitting animals and plants to meet their needs seems hard to believe. Yet the Darwinian theory is even more demanding: a single plant, a single animal would require thousands and thousands of lucky, appropriate events. Thus, miracles would become the rule: events with an infinitesimal probability could not fail to occur.... There is no law against day dreaming, but science must not indulge in it (1977, pp. 88,103).

But what of these "probabilities" of which Dr. Grassé spoke? The mathematical probability of having random mutations account for all we see around us is infinitesimal. Mutations are rare, occurring on an average of once in every ten million duplications of a DNA molecule (1 in 10^7). The problem for the evolution model is apparent because a series of **related** mutations is required. The odds of getting two mutations that are related to one another is the product of the separate probabilities (10^7 x 10^7 or 10^{14}). That is one in a hundred trillion. What about getting, say, four related mutations? The odds then become 1 in 10^{28}! Mathematician Murray Eden, one of the participants in a symposium on the mathematical probabilities of evolution, wrote:

It is our contention that if "random" is given a serious and crucial interpretation from a probabilistic point of view, the randomness postulate is highly implausible and that an adequate scientific theory of evolution must await the discovery and elucidation of new natural laws... (1967, p. 109).

Stephen J. Gould even went so far as to say: "A mutation doesn't produce major new raw material. You don't make a new species by mutating the species.... That's a common idea people have; that evolution is due to random mutations. A mutation is not the cause of evolutionary change" (1980c).

Mutations, as much as evolutionists hate to admit it, **presuppose creation**, because they are simply changes in **already existing** genes (i.e., variation within a type) that cause errors in the original template. Mutations that can cause one kind of animal to give rise to another kind of animal, or one kind of plant to give rise to another kind of plant, are unknown in the biological sciences. On the other hand, mutations that are harmful, destructive, and even lethal are known to occur.

The creation model predicts a built-in variation in the gene pool. If living things were created, variation within types is good design. Mutations, however, allegedly have introduced another kind of variation—this time harmful in nature. Mutations militate **against** evolution. And the story told regarding mutations and natural selection is much more in accord with the creation model than with the evolution model.

CONCLUSION

Carl Sagan, who was undoubtedly one of the most visible popularizers of science in our generation, once observed:

...[T]he future holds the promise that man will be able to assemble nucleotides in any desired sequence to produce whatever characteristics of human beings are thought desirable, **an awesome and disquieting prospect** (1997, 22:967, emp. added).

Yes, it is indeed an "awesome and disquieting prospect." Henry Greely, a medical bioethicist at Stanford University, commented on where this kind of thinking may lead when he wrote: "The problem is, we sanctify DNA. People seem to want to be eager to view their genome as their essence, instead of just molecules that pass on certain traits. In our secular culture, it's almost taken the place of the soul" (as quoted in Kloehn and Salopek, 1997, p. C-1).

During an interview with Stanford geneticist David Cox for the August 14, 2000 issue of *People* magazine, reporter Giovanna Breu remarked: "Some worry that mapping the genome allows us to play God by manipulating life." Dr. Cox, however, responded:

> The genome gives us a list of what living things are made up of, but not how they go together and work. It provides one more piece of information that we can start using to make order out of our ignorance and help people to make better decisions in life. But... we just have the parts, not the entire instruction manual. I think God isn't so stupid as to let anyone have that (as quoted in Breu, 2000, 54[7]:131).

While I, personally, might not have phrased my sentiments in exactly those words, it certainly is invigorating to see a scientist of Dr. Cox's stature give credit where credit is due for the creation of the "book of life" to which we refer somewhat nonchalantly as the "human genome." And it similarly is refreshing to be able to report that he is not the only scientist involved in the project who has acknowledged the Author of the intricate genetic code. At the June 26, 2000 press conference held jointly by the President of the United States and the Prime Minister of Great Britain, Dr. Francis Collins, who chairs the Human Genome Project from the National Institutes of Health, spoke in similar terms when he said:

> Today, we deliver, ahead of schedule again, the most visible and spectacular milestone of all.... We have developed a map of overlapping fragments that includes 97 percent of the human genome, and we have sequenced 85 percent of this.... It's a happy day for the world. It is humbling for me and awe-inspiring to realize that **we have caught the first glimpse of our own instruction book, previously known only to God. What a profound responsibility it is to do this work** (see Office of Technology Policy, 2000, emp. added). [NOTE: In an interview that appeared

in the March issue of *Discover* magazine three months earlier, Dr. Collins publicly affirmed his belief in an intelligent designer, and commented on how grateful he was to be associated with the HGP as it uncovered some of the "mysteries of human biology"—see Glausiusz, 2000, 21[3]:22.]

A profound responsibility indeed! To actually be able to "peek inside" the biochemical genetic code is indeed "humbling and awe-inspiring." And—regardless of how deep we probe or how intelligent we think we are—may it ever be so!

7

THE LAWS OF
PROBABILITY

One of the limitations of science is that, by its very nature, it deals not with absolute proof, but with probability. In the widely used biology text that he co-authored, George Gaylord Simpson warned the student of this fact when he said:

> We speak in terms of "acceptance," "confidence," and "probability," not "proof." If by proof is meant the establishment of eternal and absolute truth, open to no possible exception or modification, then proof has no place in the natural sciences. Alternatively, proof in a natural science, such as biology, must be defined as the attainment of a high degree of confidence (1965, p. 16).

Certainly, all practicing scientists would agree with Dr. Simpson. Science, because of its dependence upon the process of induction, cannot yield absolute proof. Over the years, investigators have elucidated successfully what today are known as the "laws of probability." Building upon the work of such men as Blaise Pascal, the famous French mathematician and scientist, others forged the principles that are employed today on a daily basis in almost every scientific discipline. George Gamow was one such individual (1961b). Emile Borel was another. Dr. Borel, one of the world's foremost experts on mathematical probability, formulated what scientists and mathematicians alike refer to as the basic "law of probability," which I would like to discuss in this chapter.

At the outset of any such discussion on probabilities, however, two questions arise. First, are probabilities of any practical nature? Second, are probabilities of any usefulness in the creation/evolution controversy? "Yes," says James Coppedge, a former director of probability research who has commented on why such studies are practical in nature.

> Probability is a practical concept. The uncertainties of chance affect our everyday lives. How likely is it to rain on the particular day on which you've planned to have an outdoor activity? What are the odds your airline flight will be hijacked? Is there a good chance your car will operate without major repairs if you delay trade-in for six months? What amount of cash will probably be sufficient to take along on a planned overseas trip? What is the likelihood that you will pass a certain exam in a school course without more study? (1973, p. 39).

Dr. Coppedge similarly explained that probability studies are useful in such things as calculation of insurance rates, analysis of stock market principles and/or prices, and other such items of an everyday interest to the average person. Further, the laws of probability, to use the words of R.L. Wysong, "are proven and trustworthy. The whole of science and every day practical living is based upon the reliability of the probable happening and the improbable not" (1976, p. 81). Indeed, whether most people realize it or not, our daily lives are affected by such mathematical studies, sometimes in ways we do not even know or understand.

But, are matters of probability related to the creation/evolution controversy? Indeed they are. Harold Morowitz, former professor of biophysics at Yale University and currently at George Mason University in Fairfax, Virginia, commented that:

> Often a process is so complicated or we are so ignorant of the boundary conditions, or of the laws governing the process, that we are unable to predict the result of the process in any but a statistical fashion.... Randomness is in a certain sense a consequence of the ignorance of the observer, yet randomness itself

displays certain properties which have been turned into powerful tools in the study of the behavior of systems of atoms (1970, pp. 64,65).

And, as Coppedge has suggested:

Evolution is an ideal subject in which to apply the laws of chance. As defined earlier, evolutionary doctrine denies advance planning, and has random matter-in-motion as its basic causal source. "Chance mutations" furnish the variability upon which presently accepted evolutionary thinking in America is generally founded (1973, pp. 44-45).

Thus, since probability studies deal with randomness, and since evolution, in its entirety, is built upon the very concept of randomness, it would appear that the laws of probability could shed some light on the possibility of evolution having occurred, which is why Dr. Coppedge remarked: "A central question we will be investigating is this: Do the laws of chance allow one to consider evolution as being within the realm of conceivable probability?" (p. 45).

There are two important issues that must be addressed in this section on statistical probability. The first is whether or not—according to accepted use of the laws of probability—the origin of life via evolutionary mechanisms is **statistically probable** in the first place. The second is whether or not such scenarios are **logically possible**. It is important to recognize that any event that is logically impossible is, by definition, probabilistically impossible on the face of it. Therefore, first we shall turn our attention to the question of whether the origin of life (as evolutionists postulate it to have occurred) is possible statistically, in keeping with accepted norms established by the laws of probability.

Borel's law of probability states that the occurrence of any event, where the chances are beyond one in one followed by 50 zeroes, is an event that we can state with certainty never will happen, no matter how much time is allotted and no matter how many conceivable opportunities could exist for the event to take place (1962, chapters 1 & 3; see also 1965, p. 62). Dr. Borel, ever the practical mathematician, commented that "the principles on which the calculus of probabilities is based are extremely simple and as intuitive as the reasonings which lead an accountant through his operations" (1962, p. 1). While the nonmathematicians among us might not agree, we nevertheless have an interest in the principles involved—and for good reason. As King and Read stated in their excellent work, *Pathways to Probability*:

> We are inclined to agree with P.S. Laplace who said: "We see...that the theory of probabilities is at bottom only common sense reduced to calculation; it makes us appreciate with exactitude what reasonable minds feel by a sort of instinct, often without being able to account for it" (1963, p. 130).

With this in mind, it is interesting to note from the scientific literature some of the probability estimates regarding the formation of life by purely mechanistic processes. For example, Dr. Morowitz himself estimated that the probability for the chance formation of the smallest, simplest form of living organism known is one chance in $1 \times 10^{340,000,000}$ [that is one chance out of 1 followed by 340 million zeroes] (1968, p. 99). The size of this figure is truly staggering, since there are supposed to be only approximately 10^{80} elementary particles (electrons and protons) in the whole Universe (Sagan, 1997, 22: 967).

The late Carl Sagan estimated that the chance of life evolving on any given single planet, like the Earth, is one chance in $1 \times 10^{2,000,000,000}$ [that is one chance out of 1 followed by 2 billion zeroes] (1973, p. 46). This figure is so large that it would take 6,000 books of 300 pages each just to write the number! A number this large is so infinitely beyond one followed by 50 zeroes (Borel's upper limit for such an event to occur) that it is simply mind boggling. There is, then according to Borel's law of probability, **absolutely no chance** that life could have "evolved spontaneously" on the Earth.

Consider, further, these facts (after Morris and Parker, 1987, pp. 269-273). If we assume the Universe to be 5 billion light years in radius, and assume that it is crammed with tiny particles the size of electrons, it has been estimated that conceivably 10^{130} particles could exist in the Universe. Every structure, every process, every system, every "event" in the Universe must consist of these particles, in various combinations and interchanges. If, to be extremely generous, we assume that each particle can take part in 10^{20} (that is a hundred billion billion) events **each second**, and then allow 10^{20} seconds of cosmic history (this would correspond to 3,000 billion years or 100-200 times the current maximum estimate of the age of the Universe), then the greatest conceivable number of separate events that could take place in all of space and time would be:

$$10^{130} \times 10^{20} \times 10^{20} = 10^{170} \text{ events}$$

Why is this the case? Allow Dr. Gamow to explain: "Here we have the rule of 'multiplication of probabilities,' which states that if you want several different things, you may determine the mathematical probability of getting them by multiplying the mathematical probabilities of getting the several individual ones" (1961b, p. 208). Or, as Adler has suggested: "Break the experiment down into a sequence of small steps. Count the number of possible outcomes in each step. Then multiply these numbers" (1963, pp. 58-59). In order for life to appear, one of these events (or some combination of them) must bring a number of these particles together in a system with enough order (or stored information) to enable it to make a copy of (reproduce) itself. And this system must come into being by mere chance.

The problem is, however, that any living cell or any new organ to be added to any existing animal—even the simplest imaginable replicating system—would have to contain far more stored information than represented even by such a gigantic number as 10^{170}. In fact, Marcel E. Golay, a leading information scientist, calculated the odds against such a system organizing itself as 10^{450} to 1 (1961, p. 23). Frank Salisbury set the figure at 10^{415} to 1 (1969, 1971). If we take Dr. Golay's figure, the odds against any accidental ordering of particles into a replicating system are at least 10^{450} to 1. This is true even if it is spread out over a span of time and a series of connected events. Golay calculated the figure on the assumption that it was accomplished by a series of 1,500 **successive** events, each with a generously high probability of(note that $2^{1,500} = 10^{450}$). The probability would have been **even lower** if it had to be accomplished in a **single chance event**! It is very generous, therefore, to conclude that the probability of the simplest conceivable replicating system arising by chance just once in the Universe, in all time, is:

$$\frac{10^{170}}{10^{450}} = \frac{1}{10^{280}}$$

When the probability of the occurrence of any event is smaller than one out of the number of events that could ever possibly occur—that is, as discussed above, less than 1/170—then the probability of its occurrence is considered by mathematicians to be zero. Consequently, it can be concluded that the chance origin of life is utterly impossible. Why so? Gamow, using simple coin tosses as his example, explained the reason for such a principle holding true.

> Thus whereas for 2 or 3, or even 4 tosses, the chances to have heads each time or tails each time are still quite appreciable, in 10 tosses even 90 per cent of heads or tails is very improbable. For a still larger number of tosses, say 100 or 1000, the probability curve becomes as sharp as a needle, and the chances of getting even a small deviation from fifty-fifty distribution becomes practically nil (1961b, p. 209).

Coppedge, in speaking to Gamow's point, observed that:

> Probability theory applies mainly to "long runs." If you toss a coin just a few times, the results may vary a lot from the average. As you continue the experiment, however, it levels out to almost absolute predictability. This is called the "law of large numbers." The long run serves to average out the fluctuations that you may get in a short series. These variations are "swamped" by the long-haul average. When a large number of tries is involved, the law of averages can be depended upon quite closely. This rule, once called the "law of great numbers," is of central importance in this field

of probability. By the way, in the popular sense, probability theory, the laws of chance, and the science of probability can be considered to be simply different expressions for the same general subject (1973, pp. 47-48).

Henry Morris, in the section he authored for *What Is Creation Science?*, wrote:

> The objection is sometimes posed that, even if the probability of a living system is 10^{-280}, every other specific combination of particles might also have a similar probability of occurrence, so that one is just as likely as another. There even may be other combinations than the one with which we are familiar on earth that might turn out to be living. Such a statement overlooks the fact that, in any group of particles, there are many more meaningless combinations than ordered combinations. For example, if a system has four components connected linearly, only two (1-2-3-4, 4-3-2-1) of the 24 possible combinations possess really meaningful order. The ratio rapidly decreases as the number of components increases. The more complex and orderly a system is, the more unique it is among its possible competitors. This objection, therefore misses the point. In the example cited above, only one combination would work. There would be 10^{280} that **would not work** (1987, pp. 272-273, emp. added).

Other writers have made the same point. Wysong, for example, concluded:

> When trying to determine whether the desired results will happen, always consider that the fractions used in probabilities carry two stories with them. One tells you the chance of something happening, and the other tells you the chance that that same event will not happen; i.e., if the odds are one in ten (10%) that a certain event will occur, then likewise the odds are nine to ten (90%) that it will not. Who could reasonably believe that a coin will turn up heads 100 times in succession, when the odds for it happening are:
>
> $$\frac{1}{1,000,000,000,000,000,000,000,000,000,000} =$$
>
> (.000000000000000000000000000001%)

and the probability that it won't is:

$$\frac{999,999,999,999,999,999,999,999,999}{1,000,000,000,000,000,000,000,000,000,000} =$$

$$(99.9999999999999999999999999999\%)$$

The probability that the event will not happen is what we must believe if we are concerned about being realistic (1976, pp. 80-81).

It is not just the extreme improbability that causes us to doubt the chemical-evolution scenario; the ordered complexity of life causes us to doubt it even more. Comments from evolutionists already have been documented that show there is no known mechanism to account for items like the genetic code, ribosomes, etc. That being true, it is astonishing to read Carl Sagan's section on "The Origin of Life" in the *Encyclopaedia Britannica*. In discussing the bacterium *Escherichia coli*, Dr. Sagan noted that this one "simple" organism contains 1 x 10^{12} (a trillion) bits of data stored in its genes and chromosomes, and then observed that if we were to count every letter on every line on every page of every book in the world's largest library (10 million volumes), we would have approximately a trillion letters. In other words, the amount of data (information) contained in approximately 10 million volumes is contained in the genetic code of the "simple" *E. coli* bacterium! Yet we are asked to believe that this marvelous organism, with its obvious complexity, occurred through purely chance processes.

In light of Dr. Sagan's observations about *E. coli*, the comments of French zoologist Pierre-Paul Grassé bear mentioning.

Bacteria, despite their great production of intraspecific varieties, exhibit a great fidelity to their species. The bacillus *Escherichia coli*, whose mutants have been studied very carefully, is the best example. The reader will agree that it is surprising, to say the least, to want to prove evolution and to discover its mechanisms and then to choose as a material for this study a being which practically stabilized a billion years ago [by evolutionary standards–BT] (1977, p. 87).

Interesting, is it not, that a code of the complexity of the DNA/ RNA code was in existence a billion years ago as the result of chance processes, and yet an organism as "simple" as *E. coli* steadfastly reproduced the genetic code during all those years without changing? R.W. Kaplan, who spent years researching the possibility of the evolutionary origin of life, suggested that the probability of the simplest living organism being formed by chance processes was one chance in 10^{130}. He then stated: "One could conclude from this result that life could not have originated without a donor of information" (1971, p. 319).

Creationists suggest that "donor" was the Creator, and that the evolution model cannot circumvent basic laws of probability. Evolutionist Richard Dawkins once observed: "The more statistically improbable a thing is, the less we can believe that it just happened by blind chance. Superficially **the obvious alternative to chance is an intelligent Designer**" (1982, p. 130, emp. added). It is not "superficial" to teach, as creationists do, that design implies a Designer. Nor is it superficial to advocate that our beautifully ordered world hardly can be the result of "blind chance." Even evolutionists like Dawkins admit (although they do not like having to do so) that the "obvious alternative" to chance is an intelligent Designer—which is the very point creationists have been making for years.

Having addressed whether the mechanistic origin of life is **statistically possible**, let us now examine whether or not it is **logically possible**. Evolutionists are fond of churning out the gargantuan numbers seen above, and then asserting rather matter-of-factly that "anything can happen, given enough time." Their point is that, probabilistically speaking, just **one** chance implies that an event might be possible. I already have shown that this is not the case. However, what these same evolutionists forget is that **logically** such scenarios not only are improbable but impossible. Sproul, Gerstner, and Lindsley concluded:

The fact is, however, we have a no-chance chance creation. We must erase the "1" which appears above the line of the "1" followed by a large number of zeroes. What are the real chances of a universe created by chance? Not a chance. Chance is incapable of creating a single molecule, let alone an entire universe. Why not? Chance is **no thing**. It is not an entity. It has no being, no power, no force. It can effect nothing for it has no causal power within it, it has no **it**ness to be within. Chance is *nomina* [name—BT] not *res* [thing—BT]; it is a word which describes mathematical possibilities which, by a curious slip of the fallacy of ambiguity, slips into discussion as if it were a real entity with real power, indeed, supreme power, the power of creativity. To say the universe is created by chance is to say the universe is created by nothing, another version of self-creation (1984, p. 118, emp. in orig.).

These authors are not the only ones who have recognized what some of their colleagues have failed to see. Claude Tresmontant, eminent philosopher of science from the University of Paris, stated:

No theory of chance can explain the creation of the world. Before chance can send atoms whirling through infinite void, the atoms have to exist! What has to be explained is the being of the world and matter. It makes no sense to say that chance can account for the creation of being (1967, p. 46).

In an impressive scientific symposium held at the Wistar Institute in Philadelphia, mathematician Murray Eden addressed the idea that somehow random, chance processes can account for the ultimate successfulness of evolution. He said:

It is our contention that if "random" is given serious and crucial interpretation from a probabilistic point of view, the randomness postulate is highly implausible and that an adequate scientific theory of evolution must await the elucidation of new natural laws—physical, physico-chemical and biological (1967, p. 109).

It is past time that evolutionists admitted as much. When, by the admission of its supporters, the only way that a theory can be accepted and propagated is by the elucidation of completely new natural laws in the physical, chemical, and biological sciences, the logical impossibility of holding to such a theory under **present** natural laws hardly needs further comment. Evolution is just such a theory, and should therefore be rejected because it is impossible—both probabilistically and logically.

8

THE FOSSIL RECORD

The late, renowned evolutionist LeGros Clark once remarked that "...the really crucial evidence for evolution must
be provided by the paleontologist whose business it is to study
the fossil record" (1955, p. 7). Indeed, Dr. Clark was correct
in such an assessment. If there is ever to be any empirical evidence for evolution, by necessity it will have to come from
what has been called "the record of the rocks," for it is here
and here alone that the actual historical evidence of any evolutionary scenario will be found. In the past, some, not knowing the actual facts of the case, were confident that it was in
"nature's museum" where the evolutionist ultimately would
make his final and unassailable stand against creation. As it
turns out, however, some of the strongest evidence for creation is to be found within the fossil record.

The fact that fossils occur, and represent the environments
in which they once lived, is not under dispute. It is the **interpretation** placed on those fossils by evolutionists that creationists
call into question. And for good reason. In his book, *Bones of
Contention*, evolutionist Roger Lewin asked in regard to the
famous Piltdown fraud:

> How is it that trained men, the greatest experts of their
> day, could look at a set of modern human bones—the
> cranial fragments—and "see" a clear simian signature
> in them; and "see" in an ape's jaw the unmistakable
> signs of humanity? The answers, inevitably, have to

do with the scientists' expectations and their effects on the interpretation of data.... Data are just as often molded to fit preferred conclusions. And the interesting question then becomes "What shapes the preference of an individual or group of researchers?" not "What is the truth?" (1987, pp. 61,68).

Philip Johnson commented in a similar vein in his book, *Darwin on Trial*: "The Darwinist approach has consistently been to find some supporting fossil evidence, claim it as 'proof' for evolution, and then ignore all the difficulties" (1991, p. 84).

For example, the methodology of the evolutionist in interpreting both the location and the importance of various fossils within the geological record is widely recognized as relying upon circular reasoning. The process begins with the assumption that life has progressed from the simple to the complex (i.e., evolution is true). On this basis, the fossils then are arranged in order from the simple to the complex. "*Voilà!*," the evolutionist says, "The sequence of fossils goes from the simple to the complex. This supports our original prediction that the fossil record should show life becoming more complex through time, and thus the fossil record proves evolution true." The end result is that an assumption (which, by definition, is unproved and unprovable) is used to "prove" evolutionary theory. This logical fallacy has not escaped the attention of even evolutionary scholars. R.R. West observed:

> Contrary to what most scientists write, the fossil record does not support the Darwinian theory of evolution because it is this theory (there are several) which we use to interpret the fossil record. By doing so, we are guilty of circular reasoning if we then say the fossil record supports this theory (1968, p. 216, parenthetical comment in orig.).

Such circular reasoning, however, cannot be accepted as a valid argument for evolution.

The point to be stressed is that the actual facts of the fossil record must be considered, without recourse to evolutionary-imposed "successions" and/or concepts of long ages. It is obvious from the preceding discussion that the fossils are very much a part of the evolutionists' story, and certainly most paleontologists are evolutionists. But this does not mean that evolutionists have exclusive rights to the fossil record. It is necessary, first, to separate scientific facts from philosophical presuppositions and, second, to make decisions based on those facts.

The question to be asked is this: Do the fossils support creation or evolution? In order to establish neo-Darwinian evolution, its proponents must be able to show intermediate or transitional forms between animals and plants in every major taxonomic subdivision. This system, first devised by the Swedish biologist Carolus Linnaeus, classifies organisms at several different levels, beginning with the broadest (kingdom), and progressively narrowing through phylum, class, order, family, genus, species, and variety. Evolutionists propose a general sequence at the phylum level beginning with single-celled organisms (e.g., bacteria), and then progressing to "simple" multicellular organisms (e.g., sponges), to mollusks (e.g., scallops), to arthropods (e.g., crabs), and then to chordates (e.g., man). On a more detailed level, say by classes of animals, the sequence may begin with cartilaginous fishes (e.g., sharks), and then progress to bony fishes, to amphibians (e.g., frogs), to reptiles (e.g., crocodiles), and then to mammals (e.g., man). Almost every biology textbook exhibits evolutionary "trees of life" that show these very sequences. Surely such dramatic but gradual changes should be witnessed in the fossil record.

Charles Darwin himself postulated that there should be "innumerable transitional links" in the fossil record. The tenth chapter of *The Origin of Species* is titled, "On the Imperfection of the Geological Record." There Darwin argued that, due to the process of natural selection, "the number of intermediate varieties, which have formerly existed, [must] be truly enormous." However, he went on to admit:

> Geology assuredly does not reveal any such finely graduated organic chain; and this, perhaps, is the most obvious and serious objection which can be argued against this theory. The explanation lies, I believe, in the extreme imperfection of the geological record (1956, pp. 292-293).

This was indeed a problem for Darwin's theory, and is still a problem for the modern version of neo-Darwinian evolution. After all, is it not a bit ridiculous to expect people to accept a scientific theory as truth when its advocates have to explain why some of the critical evidence does not even exist? It would be like a prosecuting attorney trying a murder case, and saying in his opening speech: "We know that the defendant is guilty of murder, although we cannot find a motive, the weapon, the body, or any witnesses."

It is true, of course, that the fossil record is imperfect, for some potential fossil-containing layers at certain levels in some localities may have been removed or disturbed by erosive or tectonic activities. But Darwin suggested another reason for the imperfection of the fossil record—insufficient searching. In 1859, most fossil collecting had been done in Europe and the United States. However, after more than 140 years of additional paleontological work, Darwin's defense no longer can be upheld. Evolutionary geologist T.N. George of Great Britain has stated: "There is no need to apologise any longer for the poverty of the fossil record. In some ways it has become almost unmanageably rich and discovery is outpacing integration" (1960, p. 1).

Whereas it **used** to be true that some evolutionists went to the fossil record to attempt to substantiate their theory, generally that is not the case today. Some years ago, British evolutionist Mark Ridley authored an article defending the concept of evolution as a "scientific fact," yet was forced to admit what has come to be common knowledge among those involved in the creation/evolution controversy: "No real evolutionist, whether gradualist or punctuationist, uses the fossil record as evidence in favour of the theory of evolution as opposed to special creation" (1981, 90:831).

PREDICTIONS OF THE TWO MODELS

As the evidence from the fossil record is considered, it is essential to know exactly what the evolution and creation models predict, so that the predictions can be compared to the actual data. The evolution model, on the one hand, predicts: (a) The "oldest" rocks would contain evidence of the most "primitive" forms of life capable of fossilization; (b) "Younger" rocks would exhibit more "complex" forms of life; (c) A gradual change from "simple-to-complex" would be apparent; and (d) Large numbers of transitional forms would be present (as Darwin himself, quoted above, admitted). The creation model, on the other hand, predicts: (a) The "oldest" rocks would not always contain evidence of the most "primitive" forms of life, and "younger" rocks would not always contain evidence of more "complex" forms of life; (b) A "simple-to-complex" gradation of life forms would not always appear; instead, there would be a sudden "explosion" of diverse and highly complex forms of life; and (c) There would be a regular and systematic absence of transitional forms, since there were no transitional forms.

As one examines the predictions of each of the two models in light of the actual data, it becomes clear that the evidence from the fossil record is strongly **against** evolution and **for** creation, which explains why a scientist like Dr. Ridley would suggest that evolutionists no longer use the fossil record as proof of evolution. First, consider the predictions of the evolution model that the fossil record should reveal a simple-to-complex gradation of life forms. Until recently, an examination of the Precambrian strata of the geologic timetable showed no undisputed evidence of multicellular fossil forms, while the Cambrian layer (the next layer in succession) exhibited a sudden "explosion" of life forms. In years gone by, this was a serious and fundamental problem in evolutionary theory. Today evolutionists suggest that they have found, in the Precambrian era, multicellular animals that possessed neither shells nor skeletons. Labeled the "Ediacaran fossil complex," these finds include animals resembling creatures as jellyfishes, segmented worms, and possible relatives of corals, according to evolutionists. But even with these new finds, the serious, fundamental problem for evolutionists still remains. Geneticist John Klotz has explained why.

All of the animal phyla are represented in the Cambrian period except two minor soft-bodied phyla (which may have been present without leaving any fossil evidence), and the chordates. Even the chordates may have been present, since an object which looks like a fish has been discovered in Cambrian rock. It is hardly conceivable that all these forms should have originated in this period; and yet there is no evidence for the existence of many of them prior to the Cambrian period (1972, pp. 193-194).

Since Dr. Klotz's book was published, the chordates have, in fact, been found in Cambrian rocks (see Repetski, 1978, pp. 529-531). The problem of the "missing ancestors" in Precambrian rocks is as severe as it ever was. As one science text commented:

Even theoretically, to make the vast biological leap from primitive organisms to the Cambrian fauna poses enormous problems. A remarkable series of transformations is required to change a single-celled protozoan into a complex animal such as a lobster, crab, or shrimp. The new life-forms appearing in the Cambrian were not simply a cluster of similar cells; they were complex, fully formed animals with many specialized types of cells.... The new Cambrian animals represented an astonishing leap to a higher level of specialization, organization, and integration (American Scientific Affiliation, 1986, pp. 35,37).

We are being asked by evolutionists to believe that from such "ancestors" as those found in the Ediacaran complex, **all** of the major animal phyla "evolved" in the time period represented by a jump between the Precambrian and the Cambrian periods. Such is not only impossible, but also unreasonable.

Writing under the title of "When Earth Tipped, Life Went Wild" in *Science News*, Richard Monastersky remarked:

Before the Cambrian period, almost all life was microscopic, except for some enigmatic soft-bodied organisms. At the start of the Cambrian, about 544 million years ago, animals burst forth in a rash of evolutionary activity never since equaled. Ocean creatures acquired the ability to grow hard shells, and a broad range of new body plans emerged within the geologically short span of 10 million years. Paleontologists have proposed many theories to explain this revolution but have agreed on none (1997, 152:52).

Stefan Bengtson, of the Institute of Paleontology, Uppsala University, Sweden, suggested:

> If any event in life's history resembles man's creation myths, it is this sudden diversification of marine life when multicellular organisms took over as the dominant actors in ecology and evolution. Baffling (and embarrassing) to Darwin, this event still dazzles us and stands as a major biological revolution on a par with the invention of self-replication and the origin of the eukaryotic cell. The animal phyla emerged out of the Precambrian mists with most of the attributes of their modern descendants (1990, 345:765, parenthetical item in orig.).

Evolutionist Richard Dawkins of Oxford University, wrote:

> The Cambrian strata of rocks, vintage about 600 million years [evolutionists are now dating the beginning of the Cambrian at about 530 million years], are the oldest in which we find most of the major invertebrate groups. **And we find many of them already in an advanced state of evolution, the very first time they appear. It is as though they were just planted there, without any evolutionary history.** Needless to say, this appearance of sudden planting has delighted creationists (1986, p. 229, bracketed comment in orig., emp. added).

Indeed it has. In an article appearing in *American Scientist* on "The Origin of Animal Body Plans," Erwin Douglas and his colleagues discussed what Dawkins referred to as an "advanced state of evolution."

> All of the basic architectures of animals were apparently established by the close of the Cambrian explosion; subsequent evolutionary changes, even those that allowed animals to move out of the sea onto land, involved only modifications of those basic body plans. About 37 distinct body architectures are recognized among present-day animals and from the basis of the

taxonomic classification level of phyla.... Clearly, many difficult questions remain about the early radiation of animals. Why did no many unusual morphologies appear when they did, and not earlier or later? The trigger of the Cambrian explosion is still uncertain, although ideas abound (1997, 85:126,127).

As Stephen J. Gould observed: "Even the most cautious opinion holds that 500 million subsequent years of opportunity have not expanded the Cambrian range, achieved in just five million years. **The Cambrian explosion was the most remarkable and puzzling event in the history of life**" (1994, 271:86, emp. added). Or, as Andrew H. Knoll put it: "We now know that the Ediacaran radiation was indeed abrupt and that the geologic floor to the animal fossil record is both real and sharp" (1991, 265:64).

Jeffrey S. Levinton, chairman of the department of ecology and evolution at the State University of New York at Stony Brook, admitted:

> Most of evolution's dramatic leaps occurred rather abruptly and soon after multicellular organisms first evolved, nearly 600 million years ago during a period called the Cambrian. **The body plans that evolved in the Cambrian by and large served as the blueprints for those seen today. Few new major body plans have appeared since that time. Just as all automobiles are fundamentally modeled after the first four-wheel vehicles, all the evolutionary changes since the Cambrian period have been mere variations on those basic themes** (1992, 267:84, emp. added).

It is gratifying to see that "variation on basic themes"—one of the hallmarks of creation—finally has been recognized by some within the evolutionary community.

Second, if the fossil record is to offer support for evolution, it must demonstrate an unambiguous sequence of fully functional intermediate forms. By "unambiguous" and "functional" it is meant that certain conditions must be met before an organism (fossil or living) can be considered to be a true intermediate form. Henry Morris has noted that true intermediates should show:

(1) transitional or incipient structures, such as half-scales/half-feathers on reptile/birds;(2) series of gradually changing intermediates from one major kind to another, rather than sharp changes;(3) correlation of even sharp changes with geological time sequences (1982a, p. 28).

The first of these conditions for transitional forms is not satisfied by the fossil record. For instance, mammals take many forms, but all are equally mammalian; birds vary greatly, but all are avian. Further, all mammals are equally separated in their distinct features from all non-mammalian forms, and all birds are recognizably and fundamentally different from all nonavian species—the boundaries are **that** clear. Proper transitional or incipient structures never are found. The reason for this is the obvious design that is inherent in any living thing, whether it be a bacterium or a whale, a fungus or an orchid. The parts of an organism operate together in such a wonderful, functional way that to change a single component in one organ or body system would destroy the whole mechanism. Harvard paleontologist Stephen J. Gould wrote:

> The absence of fossil evidence for intermediary stages between major transitions in organic design, indeed our inability, even in our imagination, to construct functional intermediates in many cases, has been a persistent and nagging problem for gradualistic accounts of evolution (1980b, 6[1]:127).

Whenever a new species is discovered (whether fossil or living), it either fits perfectly into well-known modern groups, or is highly specialized, belonging to its own unique group and having no relationship (evolutionary or otherwise) to any other plant or animal type. This degree of consistency presents problems in finding an evolutionary sequence that will satisfy requirement number two above. The classic example of the first case (i.e., a fossil fitting nicely into already-existing taxonomic groups) is the extinct *Archaeopteryx*, which formerly was hailed as the "missing link" in the alleged scenario of reptile-to-bird evolution. Although the soft parts do not remain, and certain skeletal features are similar to some reptiles, its feathers and wings are fully formed for the act of powered flight. A study of the brain form from the inner crania of the specimen also suggests that it should be classified as completely avian, and not a reptile-like bird or a bird-like reptile (see Jerison, 1968, pp. 1381-1382; for further details on *Archaeopteryx*, see Gish, 1985, pp. 110-117; Harrub and Thompson, 2001a, 2001b).

The search for transitional forms also has been carried out among living creatures, but once again, supposed links were discovered to be distinct species or phyla. The lungfish would be a classic example of a postulated living link, being a supposed intermediate form between fish and amphibia. It has gills and fins like fish, but lungs and a heart like an amphibian. However, the lungfish's gills are fully fish-like, and its heart is fully amphibian—the individual organ systems are not in any way transitional.

In regard to the second case (i.e., an organism so unique it fits into no existing taxonomic group), peculiar organisms have been discovered in the ocean depths, or in various fossil-rich strata, but all of them were previously unknown forms. A few are so distinctly different, and hence unrelated to other species in any evolutionary sense, that whole new phyla had to be created to classify them.

Gould found the gaps in the fossil record interesting because he views them as supporting evidence for the theory of "punctuated equilibrium" that he and Niles Eldredge suggested as an explanation as to why the fossil record contains a paucity of transitional forms (see Thompson, 1989). However, the following statement from a scientist of his caliber is a substantial indictment of neo-Darwinian evolution. Gould opined: "All paleontologists know that the fossil record contains precious little in the way of intermediate forms; transitions between major groups are characteristically abrupt" (1977a, p. 24). Thus, the gaps of the fossil record exist because the transitional forms needed to fill them have not been found. These gaps are very real indeed—too real to deny or to explain away.

The third requirement for transitional forms also is not satisfied by the fossil record, because when certain organisms appear in the fossil record, they seem totally adapted to their environment and completely conformed to their distinct type. Bats, for example, appear suddenly in the fossil record 60 million years ago (according to evolutionary timetables), yet were not preceded by any known transitional forms; nor do they differ greatly from the modern species. This is only one of many exceptions. To say that the overall trend of the fossils from simple-to-complex proves evolution would be like saying that the stronger and tougher a person gets, the more successful he will become. Any truth in this analogy may apply on rare occasions in certain areas, like in the jungle, or in a boxing ring or wrestling arena, but what about the modestly built millionaire?

And, who determines "success" anyway? One can say with certainty that the complete opposite of the established evolutionary assumptions can be, and has been, demonstrated. For instance, the so-called "Cambrian explosion" (allegedly beginning about 600 million years ago) shows the sudden appearance of representatives of every major invertebrate phylum, each highly characteristic of its class, and none with preceding transitional forms. The same occurs with vertebrate types in the early-to-mid-Paleozoic (supposedly 400 million years ago), and the flowering plants (angiosperms) in the Cretaceous (137 million years ago by evolutionary estimates). Note these words from paleontologist George Gaylord Simpson as long ago as 1953: "Most new species, genera, and families, and nearly all categories above the level of families, appear in the records suddenly, and are not led up to by known, gradual, completely transitional sequences" (p. 360).

The situation has worsened appreciably since Dr. Simpson made his initial observation. For example, Gish and his co-authors have commented:

> None of the intermediate fossils that would be expected on the basis of the evolution model has been found between single-celled organisms and inverte-

brates, between invertebrates and vertebrates, between fish and amphibians, between amphibians and reptiles, between reptiles and birds or mammals, or between "lower" mammals and primates (1981, p. iv).

Perhaps this is what Michael Denton meant when he wrote in his book, *Evolution: A Theory in Crisis:*

> It is still, as it was in Darwin's day, overwhelmingly true that the first representatives of all the major classes of organisms known to biology are already highly characteristic of their class when they make their initial appearance in the fossil record. This phenomenon is particularly obvious in the case of the invertebrate fossil record. At its first appearance in the ancient paleozoic seas, invertebrate life was already divided into practically all the major groups with which we are familiar today.... Robert Barnes summed up the current situation: "...the fossil record tells us almost nothing about the evolutionary origin of phyla and classes. Intermediate forms are non-existent, undiscovered, or not recognized" (1985, pp. 162-163).

Eleven years earlier, writing in the journal, *Evolution*, David Kitts had reminded his colleagues of the very same point:

> Despite the bright promise that paleontology provides a means of "seeing" evolution, it has presented some nasty difficulties for evolutionists, the most notorious of which is the presence of "gaps" in the fossil record. **Evolution requires intermediate forms between species, and paleontology does not provide them** (1974, p. 466, emp. added).

Dr. Gould went even farther when he suggested that "the extreme rarity of transitional forms in the fossil record persists as the trade secret of paleontology. The evolutionary trees that adorn our textbooks have data only at the tips and nodes of their branches: the rest is inference, however reasonable, not the evidence of the fossils" (1977b, p. 13). And he listed two characteristics of the fossil record that cannot be ignored:

> (1) Stasis: Most species exhibit no directional change during their tenure on earth. They appear in the fossil record looking much the same as when they dis-

appear.... (2) Sudden appearance: in any local area, a species does not rise gradually by the steady transformation of its ancestors; **it appears all at once and "fully formed"** (1977b, p. 13, emp. added).

Some might think that evolutionists like Simpson, Gould and Kitts are alone in their thinking, or are addressing "anomalies." Not so. In 1978, the late paleontologist Colin Patterson, who at that time was serving as editor of the professional journal published by the British Museum of Natural History in London, and who was one of the twentieth century's foremost authorities on evolution and the fossil record, authored a book titled *Evolution*. In that volume, he spent a mere six pages or so dealing with the fossil record (and much of that was graphs and charts). On March 5, 1979, Luther Sunderland of New York wrote Dr. Patterson a letter, inquiring about this matter (and others). Dr. Patterson's response of April 10, 1979, was printed in the August 1981 issue of the *Bible-Science Newsletter*. I have in my possession a photocopy of Dr. Patterson's original letter (on the official stationery of the British Museum), in which he said, among other things:

> ...I fully agree with your comments on the lack of direct illustration of evolutionary transitions in my book. **If I knew of any, fossil or living, I would certainly have included them**.... Yet Gould and the American Museum people are hard to contradict when they say **there are no transitional fossils**.... I will lay it on the line–**there is not one such fossil for which one could make a watertight argument** (1979, emp. added).

This is the same Colin Patterson who said in a British Broadcasting Corporation television program:

> ...We have access to the tips of a tree; the tree itself is theory and people who pretend to know about the tree and to describe what went on with it, how the branches came off and the twigs came off are, I think, telling stories (1982).

The creation model **predicts** a sudden "explosion" of life–fully formed plants and animals. The creation model **predicts** a mixture of life forms. The creation model **predicts** a systematic absence of transitional forms. The actual evidence from the fossil record clearly shows: (a) fully formed life appearing suddenly; (b) a mixture of life forms (e.g., almost all, if not all, of the phyla in the Cambrian period); and (c) a serious lack of transitional forms. In his 1976 presidential address before the British Geological Association, Derek V. Ager stated:

> It must be significant that nearly all the evolutionary stories I learned as a student...have now been debunked.... The point emerges that, if we examine the fossil record in detail, whether at the level of orders or of species, we find—over and over again—not gradual evolution, but the sudden explosion of one group at the expense of another (1976, pp. 132-133).

This "sudden explosion" is verified throughout the fossil record. But there is more to it than that. As Eldredge correctly noted: "We have been looking at the fossil record as a general test of the notion that life has evolved: **to falsify that general idea, we would have to show that forms of life we considered more advanced appear earlier than the simpler forms**" (1982, p. 46).

That brings to mind the lowly trilobite, an extinct marine arthropod that once inhabited ocean bottoms and that has been designated as an "index fossil" for the Cambrian period (450-500 million years ago, according to the manner in which evolutionists date such things). Trilobites ranged in size from a fraction of an inch to 2 feet in length. Their segmented bodies were divided into a head, an abdomen, and a tail, with the head sporting compound eyes and antennae. Despite this amazing level of organization, many evolutionists consider trilobites a very primitive sort of animal.

However, I hardly can think of any example of a form of life we consider (to use Dr. Eldredge's words) "more advanced" in certain respects than the trilobite. In fact, one part of this creature in particular poses a tremendous problem for evolutionary theory. Each trilobite eye possessed a large lens made out of a mineral called calcite. This means the lens was not flexible, and thus it could not adjust for focusing like the lens in our eyes. To compensate for this, the trilobite lens incorporated no less than four complex optical principles in a system known as an "optical doublet," perhaps making it one of the most sophisticated visual systems known in the biological world. This is amazing for an animal that supposedly died out millions of years before "advanced" eyes like ours first appeared.

A number of years ago, a professional scientific journal, *The Sciences* (which is the official organ of the New York Academy of Sciences), published an article titled "Nature's Most Perfect Eye." But, surprisingly, it was not an article on the eye of the human; rather, it was an article on the eye of the trilobite! Why so? As it turns out, the trilobite possessed, to quote from *Science News*, "the most sophisticated eye lenses ever produced by nature" (Shawver, 1974, 105:72).

Why is this the case? Riccardo Levi-Setti, a professor at the University of Chicago, one of the world's experts on the trilobites, and the author of the classic scientific text that bears their name (*Trilobites*), put it like this:

> In fact, this optical doublet is a device so typically associated with human invention that its discovery in trilobites comes as something of a shock. The realization that trilobites developed and used such devices half a billion years ago makes the shock even greater. And a final discovery—that the refracting interface between the two lens elements in a trilobite's eye was designed in accordance with optical constructions worked out by Descartes and Huygens in the mid-seventeenth century—borders on sheer science fiction.... **The design of the trilobite's eye lens could well qualify for a patent disclosure** (1993, pp. 54,57, emp. added).

Dr. Eldredge admitted that to falsify evolution, "we would have to show that forms of life we considered more advanced appear earlier than the simpler forms." Exactly—task completed! Trilobites **are** far more advanced, and **do** appear much earlier, than numerous "simpler" forms. And that is something from the fossil record that evolution cannot begin to explain.

Consider also the story from "the record of the rocks" as told by an unusual variety of additional fossils. Embedded in sedimentary rocks all over the globe are what are known as "polystrate" (or polystratic) fossils. [N.A. Rupke, a young geologist from the State University of Groningen in the Netherlands, first coined the term "polystrate fossils" (see Morris, 1970, p. 102).] Polystrate means "many layers," and refers to fossils that cut through at least two sedimentary-rock layers. Henry Morris discussed polystrate fossils in his book, *Biblical Cosmology and Modern Science,* where he first explained the process of stratification.

> Stratification (or layered sequence) is a universal characteristic of sedimentary rocks. A stratum of sediment is formed by deposition under essentially continuous and uniform hydraulic conditions. When the sedimentation stops for a while before another period of deposition, the new stratum will be visibly distinguishable from the earlier by a stratification line (actually a surface). Distinct strata also result when there is a change in the velocity of flow or other hydraulic characteristics. Sedimentary beds as now found are typically composed of many "strata," and it is in such beds that most fossils are found (1970, p. 101, parenthetical items in orig.).

Morris then went on to explain that "large fossils...are found which extend through several strata, often 20 feet or more in thickness" (p. 102). Ken Ham has noted: "There are a number of places on the earth where fossils actually penetrate more than one layer of rock. These are called 'polystrate fossils'" (2000, p. 138). Such phenomena clearly violate the idea of a gradually accumulated geologic column since, generally speaking, an evolutionary overview of that column suggests that each stratum (layer) was laid down over many thousands (or even millions) of years. Yet as Scott M. Huse remarked in his book, *The Collapse of Evolution*:

> Polystratic trees are fossil trees that extend through several layers of strata, often twenty feet or more in length. There is no doubt that this type of fossil was formed relatively quickly; otherwise it would have decomposed while waiting for strata to slowly accumulate around it (1997, p. 96).

Probably the most widely recognized of the polystrate fossils are tree trunks that extend vertically through two, three, four, or more sections of rock–rock that supposedly was deposited during vast epochs of time. However, organic material (such as wood) that is exposed to the elements will rot, not fossilize. Thus, the entire length of these tree trunks must have been preserved quickly, which suggests that the sedimentary layers surrounding them must have been deposited rapidly–possibly (and likely) during a single catastrophe (see Ham, 2000, p. 138). As Leonard Brand explained, even if the trees had been removed from oxygen, "anaerobic bacteria cause decay unless the specimens are buried rapidly" (1997, p. 240). Consequently, it is irrational to conclude from such evidence that these formations built up slowly over millions of years.

The logical explanation for such formations is that they must have been formed quickly under cataclysmic conditions. Ken Ham stated:

> For example, at the Joggins, in Nova Scotia, there are many erect fossil trees that are scattered throughout 2,500 feet of layers. You can actually see these fossil trees, which are beautifully preserved, penetrate through layers that were supposedly laid down over millions of years (p. 138).

In what surely must be a classic case of understatement, Rupke wrote concerning the Joggins polystrate fossils: "Only a wholly uncommon process of sedimentation can account for conditions like these" (1973, p. 154). [For reviews of the Joggins polystrate fossils, see Rupke, 1973, p. 154; Corliss, 1990, pp. 254-256;.] In other words, these erect fossil trees must have required a speedy burial in order to be preserved. What better evidence for a catastrophic event than trees fossilized in an **upright** position and traversing **multiple layers** of the geologic column? As Paul Ackerman correctly observed, the polystratic tree trunks "constitute a sort of frozen time clock from the past, indicating that terrible things occurred–not over millions of years but very quickly" (1986, p. 84).

This type of phenomenon is not an isolated one. Rupke produced a photograph of "a lofty trunk, exposed in a sandstone quarry near Edinburgh [Scotland], which measured no less than 25 meters and, intersecting 10 or 12 different strata, leaned at an angle of about 40°" (1973, p. 154). Thus, this particular tree must have been buried **while falling down**! In fact, one scientist who examined the tree, George Fairholme, commented on the fact that an inclined trunk constitutes a much stronger testimony for rapidity in deposition than an upright one because

> ...while the latter might be supposed to have been capable of retaining an upright position, in a semi-fluid mass, for a long time, by the mere laws of gravity, the other must, by the very same laws, have fallen, from its inclined to a horizontal position, had it not be **retained in its inclined position by the rapid accumulation of its present stony matrix** (1837, p. 394, emp. added).

In his book, *The Creation-Evolution Controversy*, R.L. Wysong presented a photograph of another extremely unusual polystrate tree. The caption underneath the photograph read as follows:

> This fossil tree penetrates a visible distance of ten feet through volcanic sandstone of the Clarno formation in Oregon. Potassium-Argon dating of the nearby John Day formation suggests that 1,000 feet of rock was deposited over a period of about seven million years or, in other words, at the rate of the thickness of this page annually! However, catastrophic burial must have formed the rock and caused the fossilization, otherwise the tree would have rotted and collapsed (1976, p. 366; see Nevins, 1974, 10[4]:191-207 for additional details).

After discussing the effects of the May 1980 eruption of Mount St. Helens, geologist Trevor Major remarked: "[U]pright tree stumps found in many coal beds represent, not the remains of trees growing in a peat swamp, but the effects of a flood or similar disaster" (1996, p. 16). William J. Fritz, an evolutionist, recognized the phenomenon in fossilized trees at Yellowstone and stated: "I do not think that entire Eocene forests were preserved *in situ* [in place–BT] even though some upright trees apparently **were preserved where they grew**" (1980a, p. 313, emp. added). In another article from the same year and same scientific journal, Fritz wrote:

> Deposits of recent mud flows on Mount St. Helens demonstrate conclusively that stumps can be transported and deposited upright. These observations support conclusions that some vertical trees in the Yellowstone "fossil forests" were transported in a **geologic situation directly comparable to that of Mount St. Helens** (1980b, p. 588, emp. added).

Fritz has acknowledged that the fossil forests at Yellowstone might have been transported by a...catastrophe! Evolutionary uniformitarianism would have us believe that the same processes going on in nature today have formed the Earth–as opposed to large-scale catastrophes and disasters. However, in light of the evidence from polystrate fossils, creationists would suggest that just the opposite is true. Some scientists have suggested that the fossil forests in Yellowstone were transported by geologic activity such as a flood and/or volcanic activity (see Brand, 1997, p. 69; Roth, 1998, p. 246). What better way to explain a marvel like Yellowstone than via such catastrophes?

Furthermore, as Henry Morris and Gary Parker discussed in their text, *What is Creation Science?*: "Polystrates are especially common in coal formations. For years and years, students have been taught that coal represents the remains of swamp plants slowly accumulated as peat and then even more slowly changed into coal" (1987, p. 168). If polystrate fossils must form quickly in order to be preserved, and if (as many evolutionists believe) coal has been formed over periods lasting millions of years, how could there be so many (or any!) polystrate fossils in coal veins? The answer, of course, is that the evolutionary scenario requiring vast eons of time for the origin of coal (and, for that matter, oil) is wrong. Geologist Steven Austin, in fact, has been working on a new concept of coal formation that does not require such lengthy spans of time (1979). Trevor Major also has addressed the possibility of rapid formation of coal and oil in his book, *Genesis and the Origin of Coal and Oil* (1996). The truth of the matter is, neither coal nor oil formation requires millions of years, but instead can occur in relatively short periods of time. This has been documented both in nature and in the laboratory (see Major, 1996, pp. 12-15).

Yet tree trunks are not the only representatives of polystrate fossils. In the state of Oklahoma, geologist John Morris studied limestone layers that contained fossilized reed-like creatures known as Calamites, which ranged from one inch to six inches in diameter. Dr. Morris noted: "These segmented 'stems' were evidently quite fragile once dead, for they are usually found in tiny fragments. Obviously, the limestones couldn't have accumulated slowly and gradually around a still-growing organism, but must have been quite rapidly deposited in a series of underwater events" (1994, p. 101). And, at times, even animals' bodies form polystrate fossils (like catfish in the Green River Formation in Wyoming—see Morris, 1994, p. 102).

But perhaps the most famous of all animal polystrate fossils is that of a baleen whale discovered in 1976. K.M. Reese, a staff writer for the peer-reviewed scientific journal, *Chemical and Engineering News*, reported the find in great detail in the October 11, 1976 issue of that publication.

> Workers at the Dicalite division of Grefco, Inc. have found the fossil skeleton of a baleen whale some 10 to 12 million years old in the company's diatomaceous earth quarries in Lompoc, California. They've found fossils there before; in fact, the machinery operators have learned a good deal about them and carefully annotate any they find with the name of the collector, the date, and the exact place found. Each discovery is turned over to Lawrence G. Barnes at the Natural History Museum of Los Angeles County. The whale, however, is one of the largest fossils ever collected anywhere. It was spotted by operator James Darrah and Dr. Barnes is directing the excavation. The whale is standing on end in the quarry and is being exposed

gradually as the diatomite is mined. Only the head and a small part of the body are visible as yet. The modern baleen whale is 80 to 90 feet long and has a head of similar size, indicating that the fossil may be close to 80 feet long (1976, 54[4]:40).

In the January 24, 1977 issue of *Chemical and Engineering News,* Larry S. Helmick, professor of chemistry at Cedarville College in Cedarville, Ohio, wrote to the editor to comment on this unusual find, and suggested:

K.M. Reese made no comment concerning the implications of the unique discovery of a baleen whale skeleton in a vertical orientation in a diatomaceous earth quarry in Lompoc, California. However, the fact that the whale is standing on end as well as the fact that it is buried in diatomaceous earth would strongly suggest that it was buried under very unusual and rapid catastrophic conditions. The vertical orientation of the whale is also reminiscent of observations of vertical tree trunks extending through several successive coal seams. Such phenomena cannot easily be explained by uniformitarian theories, but fit readily into an historical framework based upon the recent and dynamic universal flood described in Genesis, chapters 6-9 (1977, 55[4]:5).

The amazing part of this story, however, concerns the response from the scientific community to the Reese report, and Dr. Helmick's letter to the editor about the find. Read what one scientist, Harvey Olney, wrote in a letter to the editor of *Chemical and Engineering News*–and believe it if you can.

Dr. Helmick, how dare you imply that our geology textbooks and uniformitarian theories could possibly be wrong! Everybody knows that diatomaceous earth beds are built up slowly over millions of years as diatom skeletons slowly settle out on the ocean floor. **The baleen whale simply stood on its tail for 100,000 years, its skeleton decomposing, while the diatomaceous snow covered its frame millimeter by millimeter. Certainly you wouldn't expect intelligent and informed establishment scientists of this modern age to revert to the outmoded views of our forefathers just to explain such finds!** (1977, 55[12]:4, emp. added).

There you have it. Rather than accept the straightforward facts at face value, and admit that gradualistic, uniformitarian processes simply do not work, we are expected to believe instead that a whale carcass stood on its tail—decomposing all the while—as millions of tiny diatom skeletons enshrouded it over a period of more than 100,000 years! [And to suggest otherwise is to "revert to the outmoded views of our forefathers."] Yet evolutionists have the gall to carp that creationists are the ones who are gullible and refuse to accept the scientific facts? [For an in-depth examination of the baleen whale polystrate fossil, see Snelling, 1995.]

After Dr. Rupke (who, remember, was responsible for coining the term "polystrate fossils" in the first place) had cited numerous examples of such fossils (1973, pp. 152-157), he wrote: "Nowadays, most geologists uphold a uniform process of sedimentation during the earth's history; **but their views are contradicted by plain facts**" (p. 157, emp. added). Contradicted by plain facts indeed! What caused these polystrate fossils (which are found quite literally around the world)? Rupke concluded: "Personally, I am of the opinion that the polystrate fossils constitute a crucial phenomenon both to the actuality and the mechanism of a **cataclysmal deposition**" (1973, p. 157). That, of course, is exactly what creationists have said for centuries.

HUMAN EVOLUTION

Let's be blunt about one thing. Of all the branches to be found on that infamous "evolutionary tree of life," the one leading to man should be the best documented. After all, as the most recent evolutionary arrival, pre-human fossils supposedly would have been exposed to natural decay processes for the shortest length of time, and thus should be better preserved and easier to find than any others. [Consider, for example, how many dinosaur fossils we possess, and those animals were supposed to have existed over a hundred million years before man!] In addition, since hominid fossils are of the greatest interest to man (because they are supposed to represent his past), it is safe to say that more people have been searching for them longer than for any other type of fossils. If there are any real transitional forms anywhere in the world, they should be documented most abundantly in the line leading from the first primate to modern man. Certainly, the fossils in this field have received more publicity than in any other. But exactly what does the human fossil record reveal? What is its central message? Lyall Watson, writing in *Science Digest,* put it bluntly:

> The fossils that decorate our family tree are so scarce that there are still more scientists than specimens. The remarkable fact is that all the physical evidence we have for human evolution can still be placed, with room to spare, inside a single coffin (1982, p. 44).

And relatively few "family tree" fossils have been found since that statement was made.

The public, of course, generally has no idea just how scarce, and how fragmentary (literally!), the "evidence" for human evolution actually is. Furthermore, it is practically impossible to determine which "family tree" one should accept. Richard Leakey (of the famed fossil-hunting family in Africa) has proposed one. His late mother, Mary Leakey, proposed another. Donald Johanson, while president of the Institute of Human Origins in Berkeley, California, proposed yet another. And Meave Leakey (Richard's wife) has proposed still another. At an annual meeting of the American Association for the Advancement of Science, anthropologists from all over the world descended on New York City to view hominid fossils exhibited by the American Museum of Natural History. Reporting on this exhibit, *Science News* had this to say:

> One sometimes wonders whether orangutans, chimps and gorillas ever sit around the tree, contemplating which is the closest relative of man. (And would they want to be?) Maybe they even chuckle at human scientists' machinations as they race to draw the definitive map of evolution on earth. If placed on top of one another, all these competing versions of our evo-

lutionary highways would make the Los Angeles free-
way system look like County Road 41 in Elkhart, In-
diana (see "Whose Ape Is It, Anyway?," 1984, p. 361,
parenthetical comment in orig.).

How, in light of such admissions, can evolutionary scientists
possibly defend the idea of ape/human evolution as a "scientifically proven fact"? This is not a case where science is acting in a "self-correcting" manner. Quite the opposite is true,
in fact. In this instance, scientists are looking at the exact same
fossil finds and drawing entirely different conclusions about
almost all of them!

The primate family (hominidae) supposedly consists of two
commonly accepted genera: *Australopithecus* and *Homo*. While
it is impossible to present **any** scenario of human evolution
upon which even the evolutionists themselves would agree,
currently the alleged scenario (gleaned from the evolutionists'
own writings) might appear something like this:

Aegyptopithecus zeuxis (28 million years ago) ➡ *Dryopithecus africanus* (20 million) ➡ *Ramapithecus breviros-
tris* (12-15 million) ➡ *Orrorin tugenensis* (6 million) ➡
Ardipithecus ramidus (5.8-4.4 million) ➡ *Kenyanthropus
platyops* (3.8 million years) ➡ *Australopithecus anamensis* (3.5 million) ➡ *Australopithecus afarensis* (3.4 million) ➡ *Homo habilis* (1.5 million) ➡ *Homo erectus* (2-
0.4 million) ➡ *Homo sapiens* (0.3 million-present).

Here, now, is what is wrong with all of this. *Aegyptopithecus
zeuxis* has been called by Richard Leakey "the first ape to
emerge from the Old World monkey stock" (1978, p. 52). No
controversy there; the animal is admittedly nothing more than
an ape. *Dryopithecus africanus* is (according to Leakey) "the
stock from which all modern apes evolved" (p. 56). But, as
evolutionists David Pilbeam and Elwyn Simons have pointed
out, *Dryopithecus* already was "too committed to ape-dom" to
be the progenitor of man (1971, p. 23). No controversy there;
the animal is admittedly an ape. What about *Ramapithecus*?
Thanks to additional work by Pilbeam, we now know that
Ramapithecus was not a hominid at all, but merely another
ape (1982, 295:232). No controversy there; the animal is ad-

mittedly an ape. What, then, shall we say of these three "ancestors" that form the tap root of man's family tree? We simply will say the same thing evolutionists have said: all three were nothing but apes.

The 13 fossil fragments that form *Orrorin tugenensis* (broken femurs, bits of lower jaw, and several teeth) were found in the Tugen Hills of Kenya in the fall of 2000 by Martin Pickford and Brigitte Senut of France, and have been controversial ever since. If *Orrorin* were considered to be a human ancestor, it would predate other candidates by around 2 million years. Pickford and Senut, however, in an even more drastic scenario, have suggested that **all the australopithecines**–even those considered to be our direct ancestors– should be relegated to a dead-end side branch in favor of *Orrorin*. Yet paleontologist David Begun of the University of Toronto has stated that scientists can't tell whether *Orrorin* was "on the line to humans, on the line to chimps, a common ancestor to both, or just an extinct side branch" (2001).

In 1994, Tim White and his coworkers described a new species known as *Australopithecus ramidus* (renamed a year later as *Ardipithecus ramidus*), which was dated at 4.4 million years. The August 23, 1999 issue of *Time* contained a feature article, "Up from the Apes," about the creature. When first found (and while still considered an australopithecine), morphologically this was the earliest, most ape-like australopithecine yet discovered, and therefore appeared to be a good candidate for the most distant common ancestor of the hominids. Dr. White eventually admitted, however, that *A. ramidus* no longer could be considered as a missing link because it possessed too many "chimp-like features." A year later, Meave Leakey and colleagues described the 3.5-4.2 million-year-old *Australopithecus anamensis*, a taxon that bears striking similarities to *Ardipithecus* (an admitted chimp) and *Pan* (the actual genus of the chimpanzees). In 1997, researchers discovered another *Ardipithecus*–*A. ramidus kadabba*–which was dated at 5.8-5.2 million years old. [The original *Ardipithecus ramidus* then was renamed *A. ramidus ramidus*.] Once again, *Time* ran a cover story on this alleged "missing link" (in its July 23, 2001 issue). What was it that convinced evolutionists that *kadabba* walked upright and was on the road to becoming man? A single toe bone!

Then, in the March 22, 2001 issue of *Nature*, Meave Leakey and her co-authors announced the discovery of *Kenyanthropus platyops* ("flat-faced-man of Kenya"). The authors described their finds as "a well-preserved temporal bone, two partial maxillae, isolated teeth, and most importantly a largely complete, **although distorted**, cranium" (410:433, emp. added). Leakey placed a tremendous amount of importance on the flatness of the facial features of this find, due to the widely acknowledged fact that more modern creatures supposedly possessed an admittedly flatter facial structure than their older, more ape-like alleged ancestors. This is no small problem, however, because creatures younger than *K. platyops*, and therefore closer to *Homo sapiens*, have much more pronounced, ape-like facial features. *K. platyops* was dated at 3.5-3.8 million years, and yet has a much flatter face than any other hominid that old. Thus, the evolutionary scenario seems to be moving in the wrong direction. Some have argued that *K. platyops* belongs more properly in the genus *Australopithecus*.

Australopithecus afarensis was discovered by Donald Johanson in 1974 at Hadar, Ethiopia. Dr. Johanson contends that this creature (nicknamed "Lucy") is the direct ancestor of man (see Johanson, 1981). Numerous evolutionists strongly disagree. Lord Solly Zuckerman, the famous British anatomist, published his views in his book, *Beyond the Ivory Tower*. He studied the australopithecines for more than 15 years and concluded that if man descended from an apelike ancestor, he did so without leaving a single trace in the fossil record (1970, p. 64). "But," some might say, " Zuckerman's work was done before Lucy was discovered." True, but that misses the point. Zuckerman's research—which established conclusively that the australopithecines were nothing but knuckle-walking apes—was performed on fossils **younger** (i.e., closer to man) than Lucy! If more recent finds are nothing but apes, how could an **older** specimen be "more human"? Charles Oxnard, while at the University of Chicago, reported his multivariate computer analysis, which documented that the australopithecines were nothing but knuckle-walking apes (1975, pp. 389-395). Then, in the April 1979 issue of *National Geographic*, Mary Leakey reported finding footprints–dated even older than Lucy at 3.6-3.8 million years–that she admitted were "remarkably similar to those of modern man" (p. 446). If Lucy gave rise to humans, then how could humans have existed more than 500,000 years before her in order to make such footprints? [See Lubenow, 1992, pp. 45-58 for a detailed refutation of Lucy.]

What of *Homo habilis*? J.T. Robinson and David Pilbeam have long argued that *H. habilis* is the same as *A. africanus*. Louis Leakey (Richard's father) even stated: "I submit that morphologically it is almost impossible to regard *Homo habilis* as representing a stage between *Australopithecus africanus* and *Homo erectus*" (1966, 209:1280-1281). Dr. Leakey later reported the contemporaneous existence of *Australopithecus, Homo habilis*, and *H. erectus* fossils at Olduvai Gorge (see M.D. Leakey, 1971, 3:272), which would make it impossible for one to be leading up to the other, as Lubenow explained when he wrote:

> When a creationist emphasizes that according to evolution, descendants can't be living as contemporaries with their ancestors, the evolutionist declares in a rather surprised tone, "Why, that's like saying that a parent has to die just because a child is born!" Many times I have seen audiences apparently satisfied with that analogy. But it is a very false one. In evolution, one species (or a portion of it) allegedly turns into a second, better-adapted species through mutation and natural selection. However, in the context of human reproduction, I do not turn into my children; I continue on as a totally independent entity. Furthermore, in evolution, a certain portion of a species turns into a more advanced species because that portion of the species allegedly possesses certain favorable mutations which the rest of the species does not possess. Thus the newer, more advanced group comes into direct competition with the older unchanged group and eventually eliminates

it through death.... The analogy used by evolutionists is without logic, and the problem of contemporaneousness remains....

This incontrovertible fact of the fossil record effectively falsifies the concept that *Homo erectus* evolved into *Homo sapiens* and that *Homo erectus* is our evolutionary ancestor. In reality, it falsifies the entire concept of human evolution (1992, pp. 121,127,129, parenthetical items and emp. in orig.).

And even more startling was Mary Leakey's discovery of the remains of a circular stone hut at the bottom of Bed I at Olduvai Gorge–**beneath** fossils of *H. habilis* in Bed II! Evolutionists have long attributed the deliberate manufacture of shelter only to *Homo sapiens,* yet Dr. Leakey discovered the australopithecines and *H. habilis* together with manufactured housing. As Duane Gish asked:

If *Australopithecus, Homo habilis,* and *Homo erectus* existed contemporaneously, how could one have been ancestral to another? And how could any of these creatures be ancestral to Man, when Man's artifacts are found at a lower stratigraphic level, directly underneath, and thus earlier in time to these supposed ancestors of Man? (1995, p. 271).

And what about *Homo erectus?* Examine a copy of the November 1985 issue of *National Geographic* and see if you can detect any differences between the pictures of *Homo erectus* and *Homo sapiens* (pp. 576-577). The fact is, there are no recognizable differences. As Ernst Mayr, the famed evolutionary taxonomist of Harvard remarked: "The *Homo erectus* stage is characterized by a body skeleton which, so far as we know, does not differ from that of modern man in any essential point" (1965, p. 632).

The fossil evidence for evolution (human or otherwise) simply is not there. Apes always have been apes, and humans always have been humans. Evolutionists certainly are in an embarrassing position today. They can find neither the transitional forms their theory demands, nor the mechanism to explain how the evolutionary process supposedly occurred. The available facts, however, do fit the creation model.

9

CONCLUSION

Evidences such as those marshaled under the headings above could be multiplied many times over. The point, however, is that creationists have an impressive arsenal of evidence to confirm the conclusion that the creation model better fits the available scientific facts than the evolution model. The one-sided indoctrination of students in this materialistic philosophy in the tax-supported public schools in our pluralistic, democratic society is a violation of academic and religious freedoms. Furthermore, it is poor science and poor education. To remedy this intolerable situation, creation scientists suggest that, excluding the use of the Bible or any other religious literature, the scientific evidence that can be adduced in favor of creation and evolution be presented thoroughly and fairly in public schools. Students, upon examining all the data and considering each alternative, may then weigh the implications and consequences of each position and decide for themselves which is credible and reasonable. That is good education, and good science, in the finest tradition of academic freedom. Even Darwin, in his "Introduction" to *The Origin of Species*, stated:

> I am well aware that scarcely a single point is discussed in this volume on which facts cannot be adduced often apparently leading to conclusions directly opposite to those at which I have arrived. A fair result can

be obtained only by fully stating and balancing the facts and arguments on both sides of each question... (1956, p. 18).

But many evolutionists seek to smother all challenges from within or without the scientific or educational establishment, concealing the fallacies and weaknesses of evolution and adamantly opposing a hearing for the scientific case for creation. Why is this so? There may exist two possibilities. First, it may be that evolutionists consider students too ignorant, or too illiterate, to be exposed to these competing ideas of origins. Thus, they must be "protected" and carefully indoctrinated in "correct" ideas by those who consider themselves to be the intellectually elite—the sole possessors of truth. Second, having carefully and deliberately constructed this fragile tower of hypotheses piled on hypotheses, it may be that evolutionists are aware of the fact that evolution will fare badly if exposed to an open and determined challenge from creation scientists, and that if this is done, the majority of students will accept creation as the better of the two concepts of origins. Regardless, it is urgent that students be exposed to all of the evidence so that these two alternative concepts of origins— creation and evolution—can compete freely in the marketplace of ideas.

APPENDIX A

GLOSSARY OF TERMS

[As one science writer observed, some scientific terminology "is a formidable thicket of jargon" (Ridley, 1999, p. 5). Therefore, in order to assist those who may not be familiar with biological/genetic terminology, I am providing the following glossary to accompany the material in chapter 6. The words and phrases in **bold type** within these definitions also appear in the glossary.]

Alleles–In **diploid** organisms, different forms of the same **gene** (arranged as homologous pairs, one having been donated by each parent) on the **DNA** molecule.

Amino Acids–The basic building blocks of proteins; organic compounds containing an acidic carboxyl (COOH) group, a basic amino (NH_2) group, and a distinctive side group ("R" group) that varies in each amino acid and that determines the individual chemical properties of each. Twenty common amino acids are found in **proteins**.

Autosome–Any **eukaryotic chromosome** not involved in sex determination. Autosomes constitute the vast majority of an organism's chromosomal complement.

Base–A nitrogen-containing (nitrogenous) molecule that, in combination with a **pentose sugar** and a phosphoric acid (**phosphate**) group, forms a **nucleotide**.

Chromosome–Threadlike structure into which **DNA** is organized, and on which **genes** (and other DNA) are carried. In **eukaryotes**, chromosomes reside in a membrane-bound cell nucleus; in **prokaryotes**, the chromosome consists of a single circle of naked DNA. From Greek *chromos* ("color") because colored stains originally were used to visualize chromosomes. The number of chromosomes is characteristic of a species (humans have 23 matched pairs–22 **autosome** pairs; one **sex chromosome** pair).

Codon–The basic coding unit in **DNA/RNA**; composed of a triplet of **nucleotides**.

Cytogenic Map–The visual appearance of a **chromosome** when stained and examined microscopically. Visually distinct regions ("light" and "dark" bands) give each chromosome a unique appearance; important in determination of aberrations.

Cytoplasm–The inside of a cell, excluding the **nucleus** and **organelles**, that is a matrix containing dissolved/suspended ions and other molecules necessary for life.

Diploid–The number of **chromosomes** in **somatic cells** (as opposed to **gametes**) of humans and animals. In diploid cells, each chromosome is present in duplicate (or twice the **haploid** number). Diploid cells normally are produced by **mitosis**, which does not reduce chromosome number (as in **meiosis**) but maintains the original number.

DNA–Deoxyribonucleic acid; a **nucleic acid** containing the genetic information found in most organisms and which is the main component of **chromosomes** of **eukaryotic** organisms. The DNA molecule is composed of two winding **polynucleotide** chains that form a **double helix**. Each chain is composed of individual units made of a **base** (adenine, cytosine, guanine, or thymine) linked via a **pentose sugar** (deoxyribose) to a **phosphate** molecule.

Double Helix–The structural arrangement of **DNA**, which looks something like a long ladder twisted into a coil (helix). The sides of the "ladder" are formed by a backbone of **pentose sugar** and **phosphate** molecules, and the "rungs" are composed of **nucleotide** bases joined weakly in the middle by hydrogen bonds.

Endoplasmic reticulum–A system of membranous sacs traversing the **cytoplasm** of **eukaryotic** cells. Provides transportation for delivery of synthesized **proteins** or for secretion of substances to the cell's exterior in conjunction with **Golgi bodies**.

Eukaryote–A cell characterized by membrane-bound **organelles** (such as the nucleus, ribosomes, et al.). Animals, plants, fungi, and protoctists are eukaryotic.

Gamete–A **haploid** reproductive cell (**spermatozoon** or sperm cell in the male; **oocyte** or egg cell in the female) capable of fusing with another reproductive cell during fertilization to produce a **diploid zygote**. In sexual reproduction, each gamete transmits its parental **genome** to the progeny. In humans and most animals, the male gamete often is smaller than its counterpart in the female, is motile, and is produced in large numbers. The female gamete, by contrast, is much larger, immotile, and produced in relatively small numbers.

Gene–The physical hereditary unit passed from parent to offspring. Genes are sequences of **nucleotides** or pieces of **DNA**, most of which contain information for producing a specific **protein**. Genes code for the structures and functions of an organism.

Genetic Map–A map (also known as a chromosomal or linkage map) showing the linear arrangement of a particular species' **genes** in relation to each other, rather than as specific points on each **chromosome**.

Genome–The total genetic makeup of an organism (from the Greek *génos*, "generation" or "kind"). Refers to **DNA** complement of a **haploid** cell, including DNA in the **chromosomes** as well as that within **mitochondria**. ["Nuclear genome" refers solely to DNA within the nucleus; "human genome" refers to all the DNA contained in an entire human (**haploid**) cell, rather than just in the nucleus.]

Genotype–The genetic identity of an individual that does not show as outward characteristics, but instead is a description of all **genes** that are present in the **genome** regardless of their state of expression or modification. **Phenotype** often is apparent to the naked eye; genotype can be determined only by specific genetic testing.

Germ cell–see **Gamete**.

Golgi Body–An **organelle** present in **eukaryotic** cells that functions as a collection and/or packaging center for substances that the cell manufacturers for transport. Especially useful in **protein** distribution.

Haploid–The number of chromosomes in a **spermatozoon** or **oocyte**; half the **diploid** number. Haploid cells normally are produced by **meiosis**, which reduces the **chromosome** number by half during the formation of **gametes**.

Meiosis–The ordered process of cell division by which the chromosome number is reduced by half. Meiosis is the key element in the production of **haploid gametes**.

Mitochondria–The cellular **organelles** found in **eukaryotic** cells where energy production and respiration occur.

Mitosis–The ordered process by which a cell divides to produce two identical progeny, each with the same number of **chromosomes** as the original parent cell.

Nucleic Acid–see **Polynucleotide**.

Nucleotide–One of the structural components of **DNA** and **RNA**; composed of one sugar molecule, one phosphoric acid molecule, and one nitrogenous **base** molecule (adenine, cytosine, guanine, orthymine). ["Base" and "nucleotide" are used interchangeably in referring to residues that compose **polynucleotide** chains of DNA or RNA.]

Oocyte–The mature, female reproductive cell (also known as an egg cell).

Organelle–A subcellular structure characteristic of **eukaryotic** cells that performs a specific function. Largest organelle is the nucleus; others include **Golgi bodies**, **ribosomes**, and the **endoplasmic reticulum**.

Pentose Sugar–A sugar that has five carbon atoms in each molecule [e.g., ribose (in **RNA**) or deoxyribose (in **DNA**)].

Phenotype–The external, physical appearance of an organism that includes such traits as hair color, weight, height, etc. The phenotype is determined by the interaction of genes with each other and with the environment, whereas the **genotype** is strictly genetic in orientation. Phenotypic traits (e.g., weight) are not necessarily genetic.

Phosphate–Also known as phosphoric acid; element essential to living creatures. Required for energy storage and transfer (ion state also serves as a biological buffer).

Physical Mapping–Shows specific physical location of a particular species' **genes** on each **chromosome**. Physical maps are important in searches for disease-causing genes.

Polynucleotide–Also known as a **nucleic acid**. One of the four main classes of macromolecules (**proteins, nucleic acids**, carbohydrates, lipids) found in living systems. Polynucleotides–long chains composed of nucleotide–form backbone of **DNA**, in which two polynucleotide chains interact as their nitrogenous **bases** connect to form what is known as the DNA **double helix**.

Prokaryotes–cells that possess a plasma membrane, yet lack a true nucleus and membrane-bound organelles within their cytoplasm. In prokaryotes, the **DNA** normally is found in a single, naked, circular **chromosome** (known as a genophore) that lies free in the cytoplasm.

Proteins–One of four main classes of macromolecules (in addition to **nucleic acids**, carbohydrates, and lipids) in living systems. Proteins are composed of **amino acids** and perform a wide variety of activities throughout the body.

RNA–Ribonucleic acid; a nucleic acid that functions in various forms to translate information contained in **DNA** into **proteins**. Similar in composition to DNA, in that each **polynucleotide** chain is composed of units made of a **base** (adenine, cytosine, guanine, or, in the case of RNA, uracil, rather than thymine as in DNA) linked via a **pentose sugar** (in this case, ribose rather than deoxyribose) to a **phosphate** molecule. Generally is single stranded (as opposed to DNA's **double helix**), except on occasions where it (rather than DNA) serves as the primary genetic material contained in certain double-stranded RNA viruses. Numerous forms of RNA, including messenger RNA (mRNA), transfer RNA (tRNA), and ribosomal RNA (rRNA) are responsible for carrying out a variety of different functions.

Ribosomes–The intracellular, molecular machines that carry out **protein** synthesis. Associated with **RNA** and often attached to the **endoplasmic reticulum**.

Sex Cell–see **Gamete**.

Sex Chromosomes–The **chromosomes** that determine the sex of organisms which exhibit sexual differentiation (e.g., humans, most animals, some higher plants). In humans, the X chromosome determines female genetic traits; the Y chromosome determines male traits. Since a single chromosome is inherited from each parent during reproduction, XX is female, and XY is male.

Somatic Cells–All the cells (often referred to as body cells) of a multicellular organism other than the **sex cells** (**gametes**). Somatic cells reproduce only by the process of **mitosis**; changes in such cells are not heritable, since they are not involved in germ-line reproduction as **sex cells** are.

Spermatozoon–The mature, male reproductive cell (also known as a sperm cell).

Zygote–The **diploid** cell that results from the fusion of the male and female **gametes** that will grow into the embryo, fetus, and eventually the neonate (newborn).

APPENDIX B

ARP'S ANOMALIES

astrophysicist
Halton C. Arp

Halton Arp is an astrophysicist at the Max Planck Institute for Astrophysics in Munich, Germany. He has been referred to by some of his colleagues as "the most feared astronomer on Earth" (see Kaufmann, 1981). Renowned physicist John Gribbin once wrote that "for 20 years or so" Arp has been "a thorn in the side of establishment astronomy" (1987, p. 65). Depictions such as these generally are not seen in scientific literature. What, pray tell, has Dr. Arp done to deserve such designations?

For more than three decades, Arp has compiled—through extensive observational astronomy using some of the world's finest telescopes—a sizable database of peculiar galaxies and redshift anomalies. These peculiarities and anomalies hardly are insignificant or few in number. Quite the opposite—Dr.

SEEING RED

REDSHIFTS, COSMOLOGY
AND ACADEMIC SCIENCE

by Halton Arp

QUASARS, REDSHIFTS
AND CONTROVERSIES

by Halton Arp

Arp has produced an entire "Atlas of Peculiar Galaxies" (currently administered by the California Institute of Technology—see Arp, 1966). The images that he has produced, and the implications that stem from them, have struck at the very heart of current cosmological theory.

By way of summary, Arp has discovered entities (e.g., galaxies) that exhibit one redshift value (designated as "z" in the scientific literature) that are physically associated with other entities (e.g., quasars) with entirely different redshift values. As Gribbin wryly noted: **"If a galaxy and a quasar are physically connected, but have different redshifts, something definitely is wrong**.... Arp has enough evidence that he ought to be worrying more people than actually acknowledge the significance of his findings" (p. 65, emp. added).

In a personal e-mail sent on April 17, 2003, Dr. Arp referred to what he called "the latest, very powerful evidence found in the active galaxy NGC 7603 (Arp Atlas No. 92). The high-redshift companion is attached to the arm from the active galaxy which contains two very high redshift, quasarlike objects" (see images below). It was this very enigma that two Spanish astronomers discussed in a 2002 paper in *Astronomy and Astrophysics*, in which they stated: "As far as we are aware, this is the most impressive case of a system of anomalous redshifts discovered so far" (see López-Corredoira and Gutiérrez, 2002, p. L15). Alow me to explain.

In figure A below, there are **four** objects. NGC 7603 is a spiral galaxy with a redshift value of **0.029**. Object 1 is a quasar with $z = 0.057$. Objects 2 and 3 are quasar-like objects with z values of **0.243** and **0.391** respectively. As López-Corredoira and Gutiérrez noted: "Everything points to the four objects being connected among themselves, but how to explain the different redshifts?" (p. L17). How to explain indeed? Gribbin lamented: "That strikes at the foundation

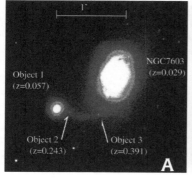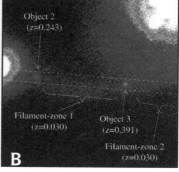

stone of received cosmological wisdom" (p. 65). It certainly does! As Dr. Arp himself put it in an e-mail to my office on March 27, 2003, this is a case where "we once again are experiencing a situation where data get thrown out if they don't fit the theory." Big Bang cosmology simply cannot explain "Arp's anomalies."

REFERENCES

Abercrombie, M., C. Hickman, and M. Johnson (1961), *A Dictionary of Biology* (Baltimore, MD: Penguin).

Ackerknect, E.H. (1973), "Rudolph Virchow," *Encyclopaedia Britannica*, 23:35.

Ackerman, Paul D. (1986), *It's a Young World After All* (Grand Rapids, MI: Baker).

Adler, Irving (1963), *Probability and Statistics for Everyman* (New York: John Day).

Ager, Derek V. (1976), "The Nature of the Fossil Record," *Proceedings of the Geological Association*, 87[2]:131-159.

Akridge, Russell, Thomas Barnes, and Harold S. Slusher (1981), "A Recent Creation Explanation of the 3 K Background Black Body Radiation," *Creation Research Society Quarterly*, 18[3]:159-162, December.

Alpher, R.A. and R. Herman (1949), *Physical Review*, 75:1089-1095.

Amato, I. (1986), "Spectral Variations on a Universal Theme," *Science News*, 130:166, September 13.

American Scientific Affiliation (1986), *Teaching Science in a Climate of Controversy* (Ipswich, MA).

Anderson, Bruce L. (1980), *Let Us Make Man* (Plainfield, NJ: Logos International).

Andrews, E.H. (1978), *From Nothing to Nature* (Welwyn, Hertfordshire, England: Evangelical Press).

Arp, Halton (1966), "Atlas of Peculiar Galaxies," [On-line], URL: http://nedwww.ipac.caltech.edu/level5/Arp/Arp_contents.html.

Arp, Halton (1987), *Quasars, Redshifts, and Controversies* (Berkeley, CA: Interstellar Media).

Arp, Halton (1999), *Seeing Red: Redshifts, Cosmology and Academic Science* (Montreal, Canada: Apeiron).

Arp, Halton (2000), "What has Science Come to?," *Journal of Scientific Explorations*, 14[3]:447-454.

Arp, H.C., G. Burbidge, F. Hoyle, J.V. Narlikar, and N.C. Wickramasinghe (1990), "The Extragalactic Universe: An Alternative View," *Nature*, 346:807-812, August 30.

Asimov, Isaac (1972), *Isaac Asimov's Biographical Encyclopedia of Science and Technology* (New York: Avon).

Augros, Robert and George Stanciu (1987), *The New Biology* (Boston, MA: New Science Library).

Austin, Steven A. (1979), *Depositional Environment of the Kentucky Number 12 Coal Bed (Middle Pennsylvanian) of Western Kentucky, with Special Reference to the Origin of Coal Lithotypes,* Unpublished Ph.D. dissertation (Pennsylvania State University).

Avers, C.J. (1989), *Process and Pattern in Evolution* (Oxford, England: Oxford University Press).

Avery, O.T., C.M. MacLeod, and M. McCarty (1944), "Studies on the Chemical Nature of the Substance Inducing Transformation of Pneumococcal Types," *Journal of Experimental Medicine*, 79:137-158.

Ayala, Francisco (1968), "Genotype, Environment, and Population Numbers," *Science*, 162:1436.

Ayala, Francisco (1978), "The Mechanisms of Evolution," *Scientific American*, 239[3]:56-69, September.

Barrow, John D. (1991), *Theories of Everything* (Oxford, England: Clarendon Press).

Barrow, John D. (2000), *The Book of Nothing: Vacuums, Voids, and the Latest Ideas about the Origins of the Universe* (New York: Pantheon).

Barrow, John D. and Frank Tipler (1986), *The Anthropic Cosmological Principle* (Oxford, England: Oxford University Press).

Bartlett, J.G., A. Blanchard, J. Silk, and M.S. Turner (1995), "The Case for a Hubble Constant of 30 kms-1 Mpc-1," *Science*, 267:980-983, February 27.

Bateson, William (1914), *Nature*, August 20.

Begun, David (2001), "Early Hominid Sows Division," [Online], URL: http://bric.postech.ac.kr/science/97now/01_2now/010222c.html.

Behe, Michael J. (1998), "Intelligent Design Theory as a Tool for Analyzing Biochemical Systems," *Mere Creation*, ed. William A. Dembski (Downers Grove, IL: InterVarsity Press).

Bengtson, Stefan (1990), "The Solution to a Jigsaw Puzzle," *Nature*, 345:765-766, June 28.

Berlinski, David (1998), "Was There a Big Bang?," *Commentary*, pp. 28-38, February.

"Big Bang Brouhaha" (1992), *Nature*, 356:731, April 30.

Bird, Wendell R. (1987), "Evolution as Theory and Conjecture: Cosmic Evolution of the Universe," *Origins Research*, Special Insert, 10[2]:I-1–I-10, Fall/Winter.

Bishop, George (1998), "The Religious Worldview and American Beliefs about Human Origins," *The Public Perspective*, August/September, pp. 39-48.

Bishop, Jerry E. and Michael Waldholz (1999), *Genome: The Story of the Most Astonishing Scientific Adventure of Our Time—The Attempt to Map All the Genes in the Human Body* (Lincoln, NE: toExcel Publishers).

Borel, Emile (1962), *Probabilities and Life* (New York: Dover).

Borel, Emile (1965), *Elements of the Theory of Probability* (Englewood Cliffs, NJ: Prentice-Hall).

Bouw, Gerardus D. (1982), "Cosmic Space and Time," *Creation Research Society Quarterly*, 19[1]:28-32, June.

Brand, Leonard (1997), *Faith, Reason, & Earth History: A Paradigm of Earth and Biological Origins by Intelligent Design* (Berrien Springs, MI: Andrews University Press).

Breu, Giovanna (2000), "The Code of Life" [Interview with geneticist David Cox, Codirector of the Human Genome Mapping Center at Stanford University], *People*, 54[7]:129-131, August 14.

Brown, Kathryn (2000), "The Human Genome Business Today," *Scientific American*, 283[1]:50-55, July.

Brumfiel, Geoff (2003), "Cosmology Gets Real," *Nature*, 422:108-110, March 13.

Cairns-Smith, A.G. (1985), *Seven Clues to the Origin of Life* (Cambridge: Cambridge University Press).

Cavalli-Sforza, Luigi (2000), *Genes, Peoples, and Languages* (New York: North Point Press).

Chaisson, E.J. (1992), "Early Results from the Hubble Space Telescope," *Scientific American*, 266[6]:44-51, June.

Check, Erika (2000), "Sound Smart About Sequencing," *Newsweek*, 136[7]:9, August 14.

Clark, LeGros (1955), *Discovery*, January.

Cline, David B. (2003), "The Search for Dark Matter," *Scientific American*, 288[3]:50-59, March.

Collins, Francis (1997), "The Human Genome Project," *Genetic Ethics: Do the Ends Justify the Genes?*, ed. John F. Kilner, Rebecca D. Pentz, and Frank E. Young (Grand Rapids, MI: Eerdmans), pp. 95-103.

Considine, Douglas M. (1976), *Van Nostrand's Scientific Encyclopedia* (New York: Van Nostrand Reinhold), fifth edition.

Coppedge, James E. (1973), *Evolution: Probable or Improbable?* (Grand Rapids, MI: Zondervan).

Corliss, William R. (1983), "Anomalous Redshifts (Again)," *Science Frontiers*, [On-line], URL: http://www.science-frontiers.com/sf027/sf027p03.htm.

Corliss, William (1990), *Neglected Geological Anomalies* (Glen Arm, MD: The Sourcebook Project).

Cousins, Norman (1985), "Commentary," in *Nobel Prize Conversations* (Dallas, TX: Saybrook). [This book is a record of conversations that occurred in November, 1982 at the Isthmus Institute in Dallas, Texas, among four Nobel laureates: Sir John Eccles, Ilya Prigogine, Roger Sperry, and Brian Josephson.]

Cowen, Ron (1990a), "Enigmas of the Sky: Partners or Strangers?," *Science News*, 137:181, March 24.

Cowen, Ron (1990b), "Galaxy Map Smooths Out the Vast Cosmos," *Science News*, 137:262, April 28.

Cowen, Ron (1991a), "Quasar Clumps Dim Cosmological Theory," *Science News*, 139:52, January 26.

Cowen, Ron (1991b), "Quasars: The Brightest and the Farthest," *Science News*, 139:276, May 4.

Cowen, Ron (1994), "Searching for Cosmology's Holy Grail," *Science News*, 146:232–234, October 8.

Cowen, Ron (2002), "Big Bang Confirmed," *Science News*, 162:195, September 28.

Cowen, Ron (2003), "Mature Before Their Time," *Science News*, 163:139-140, March 1.

Craig, William Lane (1979), *The Existence of God and the Beginning of the Universe* (San Bernardino, CA: Here's Life Publishers).

Craig, William Lane (1984), *Apologetics: An Introduction* (Chicago, IL: Moody).

Cramer, John G. (1999), "Before the Big Bang," *Analog Science Fiction & Fact Magazine*, [On-line], URL: http://www.npl.wash ington.edu/AV/altvw94/html, March.

Crézé, Michel, E. Chereul, O. Bienaymé, and C. Pichon (1998), "The Distribution of Nearby Stars in Phase Space Mapped by Hipparcos," *Astronomy and Astrophysics*, 329:920-936.

Crick, Francis, (1981), *Life Itself: Its Origin and Nature* (New York: Simon & Schuster).

"Dark-Matter Heretic" (2003), *American Scientist*, [On-line], URL: www.americanscientist.org/Issues/Sciobs03/03-01sciobmond.html.

Darwin, Charles (1956 edition), *The Origin of Species* (London: J.M. Dent & Sons).

Davies, Paul (1984), *Superforce: The Search for a Grand Unified Theory of Nature* (New York: Simon & Schuster).

Davies, Paul (1988), *The Cosmic Blueprint: New Discoveries in Nature's Creative Ability to Order the Universe* (New York: Simon & Schuster).

Davies, Paul (1992), *The Mind of God* (New York: Simon & Schuster).

Davis, George E. (1958), "Scientific Revelations Point to a God," *The Evidence of God in an Expanding Universe*, ed. John C. Monsma (New York: G.P. Putnam's Sons).

Davis, Percival and Dean Kenyon (1989), *Of Pandas and People* (Dallas, TX: Haughton).

Dawkins, Richard (1982), "The Necessity of Darwinism," *New Scientist*, 94:130-132, April 15.

Dawkins, Richard (1986), *The Blind Watchmaker* (New York: W.W. Norton).

de Bernardis, P., P.A. Ade, J.J. Bock, J.R. Bond, et al. (2000), "A Flat Universe from High-Resolution Maps of the Cosmic Microwave Background Radiation," *Nature*, 404:955-959, April 27.

Denton, Michael (1985), *Evolution: A Theory in Crisis* (London: Burnett).

Denton, Michael (1998), *Nature's Destiny: How the Laws of Biology Reveal Purpose in the Universe* (New York: Simon & Schuster).

DePree, Christopher and Alan Axelrod (2001), *The Complete Idiot's Guide to Astronomy* (Indianapolis, IN: Alpha), second edition .

DeYoung, Don B. (1995), "The Hubble Law," *Creation Ex Nihilo Technical Journal*, 9[1]:7-11.

DeYoung, Don B. (2000), "Dark Matter," *Creation Research Society Quarterly*, 36:177-194, March.

Dickerson, Richard E. (1978), "Chemical Evolution and the Origin of Life," *Scientific American*, 239[3]:70-110, September.

Donn, Jeff (1999), "Chromosome Mapped," [On-line], (ABC News), URL: http://abcnews.go.com/sections/science/DailyNews/chromosome991201.html, December 1.

Douglas, Erwin, James W. Valentine, and David Jablonski (1997), "The Origin of Animal Body Plans," *American Scientist*, 85:126-137, March/April.

Dunham, I.N. Shimizu, B.A. Roe, S. Chissoe, et al. (1999), "The DNA Sequence of Human Chromosome 22," *Nature*, 402:489-495, December 2.

Eccles, John C. (1973), *The Understanding of the Brain* (New York: McGraw-Hill).

Eccles, John C. (1984), "Modern Biology and the Turn to Belief in God," *The Intellectuals Speak Out About God*, ed. Roy A. Varghese (Chicago, IL: Regnery Gateway), pp. 47-50.

Eccles, John C. and Daniel N. Robinson (1984), *The Wonder of Our Being Human: Our Brain and Our Mind* (New York: The Free Press).

Eddington, Arthur S. (1926), *The Internal Constitution of the Stars* (Cambridge, England: Cambridge University Press).

Eden, Murray (1967), "Inadequacies of Neo-Darwinian Evolution as a Scientific Theory," *Mathematical Challenges to the Neo-Darwinian Interpretation of Evolution*, ed. Paul S. Moorhead and Martin M. Kaplan, Wistar Symposium No. 5 (Philadelphia, PA: Wistar Institute).

Edey, Maitland and Donald C. Johanson (1989), *Blueprints: Solving the Mystery of Evolution* (Boston, MA: Little, Brown).

"Editorial: Genome Sequencing for All," (2000a), *Nature*, 406:109, July 13.

"Editorial: Private vs. Public Genomics," (2000b), *Nature*, 403:117, January 13.

Efstathiou, G., W.J. Sutherland, and S.J. Maddox, (1990), "The Cosmological Constant and Cold Dark matter," *Nature*, 348:705-707, December 20.

Eiseley, Loren (1957), *The Immense Journey* (New York: Random House).

Eldredge, Niles (1982), *The Monkey Business* (New York: Pocket Books).

Estling, Ralph (1994), "The Scalp-Tinglin', Mind-Blowin', Eye-Poppin', Heart-Wrenchin', Stomach-Churnin', Foot-Stumpin', Great Big Doodley Science Show!!!," *Skeptical Inquirer*, 18[4]:428-430, Summer.

Estling, Ralph (1995), "Letter to the Editor," *Skeptical Inquirer*, 19[1]:69-70, January/February.

"Evidence for the Fine-Tuning of the Universe," (no date), [On-line], URL: http://www.godandscience.org/apologetics/designun.html.

Fairholme, George (1837), *New and Conclusive Physical Demonstrations, Both of the Fact and Period of the Mosaic Deluge, and of Its Having Been the Only Event of the Kind That Has Ever Occurred Upon the Earth* (London: T. Ridgeway & Sons).

Ferguson, Kitty (1994), *The Fire in the Equations: Science, Religion, and the Search for God* (Grand Rapids, MI: Eerdmans).

Fischer, Joannie Schrof (2000), "We've Only Just Begun," *U.S. News & World Report*, 129[1]:47, July 3.

Flam, Faye (1992), "COBE Finds the Bumps in the Big Bang," *Science*, 256:612, May 1.

Folger, Tim (1991), "Too Smooth a Universe," *Discover*, 12[1]:34-35, January.

Ford, E.B. (1979), *Understanding Genetics* (New York: Pica Press).

Fox, Karen (2002), *The Big Bang Theory–What It Is, Where It Came from, and Why It Works* (New York: John Wiley & Sons).

Fraser, C.M., J.D. Gocayne, O. White, M.D. Adams, R.A. Clayton, R.D. Fleischmann, C.J. Bult, A.R. Kerlavage, G. Sutton, J.M. Kelley, et. al. (1995), "The Minimal Gene Complement of *Mycoplasma genitalium*, *Science*, 270:397-403, October 20.

Freedman, W.L., B. Madore, J. Mould, L. Ferrarese, R. Hill, et al. (1994), "Distance to the Virgo Cluster Galaxy M100 from Hubble Space Telescope Observations of Cepheids," *Nature*, 371:757-762, October 27.

Fritz, William J. (1980a), "Reinterpretation of the Depositional Environment of the Yellowstone Fossil Forest," *Geology*, 8[7]:309-313.

Fritz, William J. (1980b), "Stumps Transported and Deposited Upright by Mount St. Helens Mud Flows," *Geology*, 8[12]: 586-588.

Gamow, George (1961a), *The Creation of the Universe* (New York: Viking).

Gamow, George (1961b), *One, Two, Three—Infinity* (New York: Viking).

Gardner, Eldon J. (1972), *The History of Biology* (Minneapolis, MN: Burgess Publishing), third edition.

Gardner, Martin (2000), *Did Adam and Eve Have Navels?* (New York: W.W. Norton).

Geisler, Norman L. (1976), *Christian Apologetics* (Grand Rapids, MI: Baker).

Geisler, Norman L. (1984), "The Collapse of Modern Atheism," *The Intellectuals Speak Out About God*, ed. Roy A. Varghese (Chicago, IL: Regnery), pp. 129-152.

Geller, Margaret J. and John P. Huchra (1989), "Mapping the Universe," *Science*, 246:897-903, November 17.

George, T.N. (1960), *Science Progress*, 48[1]:1-5, January.

Gergen, David (2000), "Collaboration? Very Cool," *U.S. News & World Report*, 129[2]:64, July 10.

Gerstner, John H. (1967), *Reasons for Faith* (Grand Rapids, MI: Baker).

Gish, Duane T. (1985), *Evolution: The Challenge of the Fossil Record* (El Cajon, CA: Master Books).

Gish, Duane T. (1995), *Evolution: The Fossils Still Say No!* (El Cajon, CA: Institute for Creation Research).

Gish, Duane T., Richard B. Bliss, and Wendell R. Bird (1981), *Summary of Scientific Evidence for Creation* [Part I], Impact #95 (El Cajon, CA: Institute for Creation Research).

Gitt, Werner (1997), *In the Beginning Was Information* (Bielefeld, Germany: Christliche Literatur-Verbreitung).

Glanz, James (1996), "Is the Dark Matter Mystery Solved?," *Science*, 372:595-596, February 2.

Glanz, James (1998), "A Dark Matter Candidate Loses Its Luster," *Science*, 281:332-333, July 17.

Glanz, James (2000), "Survey Finds Support is Strong for Teaching 2 Origin Theories," *The New York Times*, p. A-1, March 11.

Glashow, Sheldon Lee (1989), "Closing the Circle," *Discover*, 10[10]:66-72, October.

Glausiusz, Josie (2000), "Genetic Code-Breaker" [Interview with Dr. Francis Collins, Head of the U.S. Human Genome Project], *Discover*, 21[3]:22, March.

Gliedman, John (1982), "Scientists in Search of the Soul," *Science Digest*, 90[7]:77-79,105, July.

Goffeau, André (1995), "Life with 482 Genes," *Science*, 270:445-446, October 20.

Golay, Marcel J.E. (1961), "Reflections of a Communications Engineer," *Analytical Chemistry*, Vol. 33, June.

Golden, Frederic and Michael D. Lemonick (2000), "The Race Is Over," *Time*, 156[1]:19-23, July 3.

Gore, Rick (1983), "The Once and Future Universe," *National Geographic*, 163[6]:704-748, June.

Gould, Stephen J. (1977a), "The Return of Hopeful Monsters," *Natural History*, 86[6]:22-30, June/July.

Gould, Stephen J. (1977b), "Evolution's Erratic Pace," *Natural History*, 86[5]:12-16, May.

Gould, Stephen J. (1980a), "Dr. Down's Syndrome," *The Panda's Thumb* (New York: W.W. Norton), pp. 160-176.

Gould, Stephen Jay (1980b), "Is a New and General Theory of Evolution Emerging?," *Paleobiology*, 6[1]:119-130, Winter.

Gould, Stephen J. (1980c), "Is a New and General Theory of Evolution Emerging?," Hobart College speech, 2-14-80; quoted in Luther Sunderland (1984), *Darwin's Enigma* (San Diego, CA: Master Books).

Gould, Stephen J. (1994), "The Evolution of Life on Earth," *Scientific American*, 271:85-91, October.

Grassé, Pierre-Paul (1977), *The Evolution of Living Organisms* (New York: Academic Press).

Green, D.E. and R.F. Goldberger (1967), *Molecular Insights into the Living Process* (New York: Academic Press).

Greenstein, George (1988), *The Symbiotic Universe* (New York: William Morrow).

Greig, J.Y.T., ed. (1932), *Letters of David Hume* (Oxford: Oxford University Press), 1:187.

Gribbin, John (1976), "Oscillating Universe Bounces Back," *Nature*, 259:15-16.

Gribbin, John (1981), *Genesis: The Origins of Man and the Universe* (New York: Delacorte Press).

Gribbin, John (1986), "Cosmologists Move Beyond the Big Bang," *New Scientist*, 110[1511]:30.

Gribbin, John (1987), "Book Review" [of Halton Arp's *Quasars, Redshifts and Controversies*], *New Scientist*, p. 65, October.

Gribbin, John (1993), "Thumbs Up for an Older Universe," *New Scientist*, 140[1897]:14-15, October 30.

Gribbin, John (1998), *In Search of the Big Bang* (New York: Penguin), second edition.

Guillaume, C.E. (1896), *La Nature*, 24[series 2]:234.

Guth, Alan (1981), "Inflationary Universe: A Possible Solution to the Horizon and Flatness Problems," *Physical Review D*, 23:347-356.

Guth, Alan (1988), Interview in *Omni*, 11[2]:75-76,78-79,94,96-99, November.

Guth, Alan and Erick J. Weinberg (1983), "Could the Universe have Recovered from a Slow First-Order Phase Transition?," *Nuclear Physics B*, 212:321-364, February 14.

Guth, Alan and Paul Steinhardt (1984), "The Inflationary Universe," *Scientific American*, 250:116-128, May.

Haeckel, Ernst (1905), *The Wonders of Life*, trans. J. McCabe (London: Watts).

Ham, Ken (2000), *Did Adam Have a Belly Button?* (Green Forest, AR: Master Books).

Harrison, Edward (2000), *Cosmology: The Science of the Universe* (Cambridge, England: Cambridge University Press).

Harrub, Brad and Bert Thompson (2001a), "*Archaeopteryx, Archaeoraptor*, and the 'Dinosaurs-to-Birds' Theory," [Part I], *Reason & Revelation*, 21:25-31, April.

Harrub, Brad and Bert Thompson (2001b), "*Archaeopteryx, Archaeoraptor*, and the 'Dinosaurs-to-Birds' Theory," [Part II], *Reason & Revelation*, 2:33-39, May.

Hartnett, John G. (2001), "Recent Cosmic Microwave Background Data Supports Creational Cosmologies." *TJ [Technical Journal]*, 15[1]:8-12.

Harwit, M. (1973), *Astrophysical Concepts* (New York: John Wiley and Sons).

Haskins, Caryl P. (1971), "Advances and Challenges in Science in 1970," *American Scientist*, 59:298-307, May/June.

Hawking, Stephen (1988), *A Brief History of Time* (New York: Bantam).

Hawking, Stephen (1994), *Black Holes and Baby Universes* (New York: Bantam).

Hayden, Thomas (2000), "A Genome Milestone," *Newsweek*, 136[1]:51, July 3.

Heeren, Fred (1995), *Show Me God* (Wheeling, IL: Searchlight Publications).

Heidelberg, John F., Jonathan A. Elsen, William C. Nelson, et al. (2000), "DNA Sequence of Both Chromosomes of the Cholera Pathogen *Vibrio cholerae*," *Nature*, 406:477-483, August 3.

Hellemans, Alexander (1997), "Galactic Disk Contains No Dark Matter," *Science*, 278:1230, November 14.

Helmick, Larry S. (1977), "Strange Phenomena" [Letter to the Editor], *Chemical and Engineering News*, 55[4]:5, January 24.

Henig, Robin (2000), *The Monk in the Garden* (New York: Houghton-Mifflin).

Hitching, Francis (1982), *The Neck of the Giraffe* (New York: Ticknor and Fields).

Hofstadter, Douglas R. (1980), *Godel, Escher, Bach: An Eternal Golden Braid* (New York: Vintage Books).

Hooton, Ernest (1937), *Apes, Men and Morons* (New York: George Allen & Unwin).

Horgan, John (1991), "In the Beginning," *Scientific American*, 264:119, February.

Hoyle, Fred (1955), *Frontiers of Astronomy* (London: Heinemann).

Hoyle, Fred (1959), *Religion and the Scientists*, as quoted in John Barrow and Frank Tipler (1986), *The Anthropic Cosmological Principle* (Oxford, England: Oxford University Press).

Hoyle, Fred (1981a), "The Big Bang in Astronomy," *New Scientist*, 92:521-527, November 19.

Hoyle, Fred (1981b), "Hoyle on Evolution," *Nature*, 294:105, 148, November 12.

Hoyle, Fred (1982), "The Universe: Past and Present Reflections," *Annual Review of Astronomy and Astrophysics*, 20:16.

Hoyle, Fred (1984), "The Big Bang Under Attack," *Science Digest*, 92:[5]:84, May.

Hoyle, Fred and Chandra Wickramasinghe (1978), *Lifecloud* (New York: Harper & Row).

Hoyle, Fred and Chandra Wickramasinghe (1981), *Evolution from Space* (London: J.M. Dent & Sons).

Hoyle, Fred and Chandra Wickramasinghe, (1991), "Where Microbes Boldly Went," *New Scientist*, 91:415, August 13.

Hoyle, Fred, Geoffrey Burbidge, and Jayant V. Narlikar (2000), *A Different Approach to Cosmology* (Cambridge, England: Cambridge University Press).

Hubble, Edwin (1929), "A Relation Between Distance and Radial Velocity Among Extra-galactic Nebulae," *Proceedings of the National Academy of Science*, 15:168-173.

Hull, David (1974), *Philosophy of Biological Science* (Englewood Cliffs, NJ: Prentice-Hall).

"Human Genome Report Press Release" (2003), International Consortium Completes Human Genome Project, [Online], URL: http://www.ornl.gov/TechResources/Human_Genome/project/50yr.html.

Humphreys, Russell (1992), "Bumps in the Big Bang," *Impact* 233 (El Cajon, CA: Institute for Creation Research).

Huse, Scott M. (1997), *The Collapse of Evolution* (Grand Rapids, MI: Baker), third edition.

Illingworth, Valerie and John O.E. Clark (2000), *The Facts on File Dictionary of Astronomy* (New York: Checkmark Books), fourth edition.

"The Inflationary Universe" (2001), [On-line], URL: http://astsun.astro.virginia.edu/~jh8h/Foundations/chapter15.html.

Jacoby, G.H. (1994), "The Universe in Crisis," *Nature*, 371: 741-742, October 27.

Jastrow, Robert (1977), *Until the Sun Dies* (New York: W.W. Norton).

Jastrow, Robert (1978), *God and the Astronomers* (New York: W.W. Norton).

Jastrow, Robert (1982), "A Scientist Caught Between Two Faiths," Interview with Bill Durbin, *Christianity Today*, August 6.

Jeans, James (1929), *The Universe Around Us* (New York: MacMillan.

Jerison, J.H. (1968), "Brain Evolution and *Archaeopteryx*," *Nature*, pp. 1381-1382.

Johanson, Donald C. (1981), *Lucy: The Beginnings of Humankind* (New York: Simon & Schuster).

Johnson, Philip (1991), *Darwin on Trial* (Downers Grove, IL: InterVarsity Press).

Kaplan, R.W. (1971), "The Problem of Chance Information of Protobionts by Random Aggregation of Macromolecules," *Chemical Evolution and the Origin of Life*, ed. R. Buver and C. Ponnamperuma (New York: American Elsevier).

Karow, Julie (2000), "The 'Other' Genomes," *Scientific American*, 283[1]:53, July.

Kaufmann, William III (1981), "The Most Feared Astronomer on Earth," *Science Digest*, 89[6]:76-81,117, July.

Kaufmann, William III (1982), "The World's Most Controversial Astronomer," *Omega Science Digest*, pp. 74-127, February.

Kautz, Darrel (1988), *The Origin of Living Things* (Milwaukee, WI: Privately published by the author).

Kenny, Anthony (1980), *The Five Ways: St. Thomas Aquinas' Proofs of God's Existence* (South Bend, IN: University of Notre Dame Press).

Kerkut, George A. (1960), *The Implications of Evolution* (London: Pergamon).

King, A.C. and C.B. Read (1963), *Pathways to Probability* (New York: Holt, Rinehart & Winston).

Kirk, David (1975), *Biology Today* (New York: Random House).

Kitts, David G. (1974), "Paleontology and Evolutionary Theory," *Evolution*, 28:458-472, September.

Kloehn, Steve and Paul Salopek (1997), "Humanity Still at Heart, Soul of Cloning Issue: Scientists and Theologians Agree We Are Our Own Persons," *Chicago Tribune*, C-1, March 2.

Klotz, John (1972), *Genes, Genesis and Evolution* (St. Louis, MO: Concordia).

Klotz, John (1985), *Studies in Creation* (St. Louis, MO: Concordia).

Knoll, Andrew H. (1991), "End of the Proterozoic Eon," *Scientific American*, 265:64-73, October.

Koestler, Arthur (1978), *Janus: A Summing Up* (New York: Vintage Books).

Kolb, Rocky (1998), "Planting Primordial Seeds," *Astronomy*, 1998, 26[2]:42-43.

Kuhn, K. (1994), *In Quest of the Universe* (New York: West Publishing).

Lahav, Noam (1999), *Biogenesis: Theories of Life's Origins* (Oxford, England: Oxford University Press).

Lapedes D.N., ed. (1978), *McGraw-Hill Dictionary of Scientific and Technical Terms* (New York: McGraw-Hill).

Leakey, Louis S.B. (1966), "*Homo habilis, Homo erectus,* and *Australopithecus,*" *Nature,* 209:1280-1281.

Leakey, M.D. (1971), *Olduvai Gorge* (Cambridge, England: Cambridge University Press).

Leakey, Mary (1979), "Footprints in the Ashes of Time," *National Geographic,* 155[4]:446-457, April.

Leakey, Meave, et al. (2001), "New Hominin Genus from Eastern Africa Shows Diverse Middle Pliocene Lineages," *Nature,* 410:433-440, March 22.

Leakey, Richard (1978), *People of the Lake* (New York: E.P. Dutton).

Lemonick, Michael (2000), "The Genome is Mapped. Now What?," *Time,* 156[1]:24-29, July 3.

Lemonick, Michael D. (2001), "The End," *Time,* 157[25]:48-56, June 25.

Lemonick, Michael D. (2003), "Cosmic Fingerprint," *Time,* 161[8]:45, February 24.

Lerner, Eric (1991), *The Big Bang Never Happened* (New York: Vintage).

Lester, Lane P. and James C. Hefley (1998), *Human Cloning* (Grand Rapids, MI: Revell).

Lester, Lane P. and Raymond Bohlin (1984), *The Natural Limits of Biological Change* (Grand Rapids, MI: Zondervan).

Levinton, Jeffrey S. (1992), "The Big Bang of Animal Evolution," *Scientific American,* 267:84-91, November.

Levi-Setti, Riccardo (1993), *Trilobites* (Chicago, IL: University of Chicago Press).

Lewin, Roger (1987), *Bones of Contention* (New York: Simon & Schuster).

Lewontin, Richard (1978), "Adaptation," *Scientific American,* 239 [3]:212-218,220,222,228,230, September.

Lewontin, Richard (2000), *It Ain't Necessarily So: The Dream of the Human Genome and Other Illusions* (New York: New York Review of Books).

Linde, Andrei (1994), "The Self-Reproducing Inflationary Universe," *Scientific American*, 271[5]:48-55, November.

Lindley, David (1991), "Cold Dark Matter Makes an Exit," *Nature*, 349:14, January 3.

Lindsay, Jeff (2001), *The Bursting of the Big Bang*, [On-line], URL: www.jefflindsay.com/BigBang.shtml.

Lipkin, Richard (1991), "Dark Secrets of the Cosmos," *Insight*, pp. 20-23, May 13.

Lipson, H.S. (1980), "A Physicist Looks at Evolution," *Physics Bulletin*, 31:138, May.

Little, Peter (1999), "The Book of Genes," *Nature*, 402:467-468, December 2.

Livio, Mario (2000), *The Accelerating Universe* (New York: John Wiley).

López-Corredoira, Martin and Carlos M. Gutiérrez (2002), "Two Emission Line Objects with $z>0.2$ in the Optical Filament Apparently Connecting the Seyfert Galaxy NGC 7603 to Its Companion," *Astronomy and Astrophysics*, 390: L15-L18.

Løvtrup, Søren (1987), *Darwinism: The Refutation of a Myth* (London: Croom Helm).

Lubenow, Marvin (1992), *Bones of Contention* (Grand Rapids, MI: Baker).

Macbeth, Norman (1982), "Darwinism: A Time for Funerals," *Towards*, 2:18, Spring.

Macer, Darryl R.J. (1990), *Shaping Genes: Ethics, Law and Science of Using New Genetic Technology in Medicine and Agriculture* [On-line], (Eubios Ethics Institute), URL: http://www.biol.tsukuba.ac.jp/~macer/SG.html.

Macer, Darryl R.J. (2000), "Introduction to the Genome Projects," *Ethical Challenges as We Approach the End of the Human*

Genome Project, ed. Darryl R.J. Macer, [On-line], URL: http://www.biol.tsukuba.ac.jp/~macer/chgp/chgp2.html.

Macilwain, Colin (2000), "World Leaders Heap Praise on Human Genome Landmark," *Nature,* 405:983-984, June 29.

MacRobert, Alan (2003), "Turning a Corner on the New Cosmology," *Sky and Telescope,* 105[5]:16-17, May.

Maddox, John (1994), "The Genesis Code by Numbers," *Nature,* 367:111, January 13.

Major, Trevor J. (1991a), "In the News–National Beliefs Polled," *Reason & Revelation,* 11:48, December.

Major, Trevor J. (1991b), "The Big Bang in Crisis," *Reason & Revelation,* 11:21-24, June.

Major, Trevor (1996), *Genesis and the Origin of Coal & Oil* (Montgomery, AL: Apologetics Press).

Marshall, Eliot (1990), "Science Beyond the Pale," *Science,* 249:14-16, July 6.

Marshall, Eliot (2000), "Rival Genome Sequencers Celebrate a Milestone Together," *Science,* 288:2294-2295, June 30.

Martin, C.P. (1953), "A Non-Geneticist Looks at Evolution," *American Scientist,* Vol. 41.

Martin, Roy C. Jr. (1999), *Astronomy on Trial* (Lanham, MD: University Press of America).

Matthews, R. (1994), "Cosmology: Spoiling a Universal 'Fudge Factor,' " *Science,* 265:740-741, August 5.

"MAXIMA–a Balloon-borne Experiment Directed by UC Berkeley, Finds Evidence for a Flat Universe, Inflation, and a Cosmological Constant," (2000), [On-line], URL: http://www.berkeley.edu/news/media/releases/2000/05/09_maxima.html, May 9.

Mayr, Ernst (1965), *Animal Species and Evolution* (Boston, MA: Harvard University Press).

Mayr, Ernst (1997), *This is Biology* (Cambridge, MA: Belknap Press of Harvard University).

McFadden, John J. (2000), *Quantum Evolution: The New Science of Life* (New York: W.W. Norton).

McGaugh, Stacy (2000), "Boomerang Data Suggest a Purely Baryonic Universe," *Astrophysics Journal*, 541:L33-L36.

Meiklejohn, J.M.D., trans. (1878), Immanuel Kant, *Critique of Pure Reason* (London).

Mendel, Gregor (1865), *Experiments in Plant Hybridization*, reprinted in J.A. Peters, ed. (1959), *Classic Papers in Genetics* (Englewood Cliffs, NJ: Prentice-Hall).

Meyer, Stephen C. (1998), "The Explanatory Power of Design: DNA and the Origin of Information," *Mere Creation*, ed. William A. Dembski (Downers Grove, IL: InterVarsity Press).

Milgrom, Mordehai (2002), "Does Dark Matter Really Exist," *Scientific American*, 287[2]:42-52, August.

Misner, Charles W., Kip S. Thorne, and John A. Wheeler (1973), *Gravitation* (San Francisco: W.H. Freeman).

Moe, Martin A. (1981), "Genes on Ice," *Science Digest*, 89[11]: 36,95, December.

Monastersky, Richard (1997), "When Earth Tipped, Life Went Wild," *Science News*, 152:52, July 26.

Monod, Jacques (1972), *Chance and Necessity* (London: Collins).

Moore, David W. (1999), "Americans Support Teaching Creationism as Well as Evolution in Public Schools," [On-line], URL: http://www.gallup.com/poll/releasespr990830. asp (Princeton, NJ: Gallup News Service).

Moore, John A. (1981), "Countering the Creationists," a paper presented to the National Academy of Sciences *Ad Hoc* Committee on Creationism in Washington, D.C., October 19.

Moore, John N., and H.S. Slusher (1974), *Biology: A Search for Order in Complexity* (Grand Rapids, MI: Zondervan).

Morowitz, Harold J. (1970), *Entropy for Biologists* (New York: Academic Press).

Morris, Henry M. (1970), *Biblical Cosmology and Modern Science* (Grand Rapids, MI: Baker).

Morris, Henry M., ed. (1974), *Scientific Creationism* (San Diego, CA: Creation-Life Publishers).

Morris, Henry M. (1982a), *Creation and Its Critics* (San Diego, CA: Creation-Life Publishers).

Morris, Henry M. (1982b), *Evolution in Turmoil* (San Diego, CA: Creation-Life Publishers).

Morris, Henry M. (1984), *The Biblical Basis for Modern Science* (Grand Rapids, MI: Baker).

Morris, Henry M. and Gary E. Parker (1987), *What Is Creation Science?* (El Cajon, CA: Master Books).

Morris, John D. (1992), "Do Americans Believe in Creation?," *Acts & Facts*, 21[2]:d, February.

Morris, John D. (1994), *The Young Earth* (Colorado Springs, CO: Master Books).

Morrison, Philip and Phylis (2001), "The Big Bang: Wit or Wisdom?," *Scientific American*, 284[2]:93,95, February.

Muir, Hazel (2002), "Death Star," *New Scientist*, 173[2326]:26, January 19.

Mulfinger, George (1967), "Examining the Cosmogonies–A Historical Review," *Creation Research Society Quarterly*, 4[2]: 57-69, September.

Munitz, Milton K. (1957), *Theories of the Universe; from Babylonian Myth to Modern Science* (Glencoe, IL: Free Press).

Murphy, Nancey and George F.R. Ellis (1996), *On the Moral Nature of the Universe* (Minneapolis, MN: Fortress).

Murray, Michael J. (1999), *Reason for the Hope Within* (Grand Rapids, MI: Eerdmans).

Musser, George (2000), "Boomerang Effect," *Scientific American*, 283[1]:14-15, July.

Narlikar, Jayant (1981), "Was There a Big Bang?," *New Scientist*, 91:19-21, July 2.

Nevins, Stuart E. (1974), "Post-Flood Strata of the John Day Country, Northeastern Oregon," *The Creation Research Society Quarterly*, 10[4]:191-214, March.

"New Theories Dispute the Existence of Black Holes" (2002), [On-line], URL: http://www.cosmiverse.com/space0117 0204.html.

Newport, Frank (1993), "God Created Humankind, Most Believe," *Sunday Oklahoman*, A-22.

Nicholl, Desmond S.T. (1994), *An Introduction to Genetic Engineering* (Cambridge, England: Cambridge University Press).

Oard, Michael (2000), "Doppler Toppler?," *Creation Ex Nihilo Technical Journal*, 14[3]:39-45.

Office of Technology Assessment–U.S. Congress (1988), *Mapping Our Genes–The Genome Projects: How Big, How Fast?* (Washington D.C.: United States Government Printing Office).

Office of Technology Policy–The White House (2000), *Remarks by the President–The Entire Human Genome Project*, [On-line], URL: http://www.whitehouse.gov/WH/EOP/OSTPhtml/ 00628_2 .html.

Olney, Harvey O. III (1977), "A Whale of a Tale" [Letter to the Editor], *Chemical and Engineering News*, 55[12]:4, March 21.

Orgel, Leslie (1982), "Darwinism at the Very Beginning of Life," *New Scientist*, 94:149-152, April 15.

Overbye, Dennis (2001), "Before the Big Bang, There Was... What?," *The New York Times* [On-line], URL: http://www. nytimes.com/2001/05/22/science/22BANG.html.

Overman, Dean L. (1997), *A Case Against Accident and Self-Organization* (Lanham, MD: Rowman and Littlefield).

Oxnard, Charles E. (1975), "The Place of the *Australopithecines* in Human Evolution: Grounds for Doubt?," *Nature*, 258: 389-395, December.

Page, Don (1983), "Inflation does not Explain Time Asymmetry," *Nature*, 304:39-41, July 7.

Palca, Joseph (1991), "In Search of 'Dark Matter,' " *Science*, 251: 22, January 4.

Pasachoff, J. (1992), *Journey through the Universe* (New York: Saunders College Publishing).

Patterson, Colin (1979), Letter of April 10, 1979 to Luther Sunderland; reprinted in *Bible-Science Newsletter*, 19[8]:8, August 1981.

Patterson, Colin (1981), Written transcript made from audio tape of lecture presented at the American Museum of Natural History, November.

Patterson, Colin (1982), "Cladistics," Interview on British Broadcasting Corporation television program on March 4. Brian Leek, producer; Peter Franz, interviewer. A written report of this interview was published in the BBC publication, *The Listener*, 106:390.

Perloff, James (1999), *Tornado in a Junkyard* (Arlington, MA: Refuge Books).

Peseley, G.A. (1982), "The Epistemological Status of Natural Selection," *Laval Theologique et Philosophique*, 38:74, February.

Peterson, Ivars (1990), "The COBE Universe: Portrait at 300,000," *Science News*, 137:245, April 21.

Peterson, Ivars (1991), "State of the Universe: If not with a Big Bang, Then What?," *Science News*, 139:232-235, April 13.

Petit, Charles, (1998), "A Few Starry and Universal Truths," *U.S. News & World Report*, 124[2]:58, January 19.

Pilbeam, David (1982), "New Hominoid Skull Material from the Miocene of Pakistan," *Nature*, 295:232-234, January.

Pilbeam, David and Elwyn Simons (1971), "A Gorilla-Sized Ape from the Miocene of India," *Science*, 173:23, July.

Popper, Karl R. (1975), *Objective Knowledge* (Oxford: Clarendon Press).

Popper, Karl R. and John C. Eccles (1977), *The Self and Its Brain* (New York: Springer International).

Pratchett, Terry (1994), *Lords and Ladies* (New York: HarperPrism).

Preuss, Paul (2000), "MAXIMA Project's Imaging of Early Universe Agrees It Is Flat, But...," [On-line], URL: http://www.lbl.gov/Science-Articles/Archive/maxima-results.html.

Radecke, Hans-Dieter (1997), "Letter," *Science,* 275:603, January 31.

Radman, M. and R. Wagner (1988), "The High Fidelity of DNA Duplication," *Scientific American,* 259[2]:40-46, February.

Reese, K.M. (1976), "Workers Find Whale in Diatomaceous Earth Quarry," *Chemical and Engineering News,* 54[4]:40, October 11.

Regalado, Antonio (2000), "Riding the DNA Railroad," *Technology Review,* 103[4]:94-98, July/August.

Regener, Erhard (1933), *Zeitschrift fur Physik,* 106:633-661, English translation by Gabriella Moesle.

Repetski, John E. (1978), "A Fish from the Upper Cambrian of North America," *Science,* 200:529-530.

Repp, Andrew (2003), "The Nature of Redshifts and an Argument by Gentry," *Creation Research Society Quarterly,* 39 [4]:269-274, March.

Ridley, Mark (1981), "Who Doubts Evolution?" *New Scientist,* 90:830-832, June 25.

Ridley, Matt (1999), *Genome: Autobiography of a Species in 23 Chapters* (New York: HarperCollins).

Ross, Hugh (1991), *The Fingerprint of God* (Orange, CA: Promise Publishing).

Ross, Hugh (1993), *The Creator and the Cosmos* (Colorado Springs, CO: Navpress).

Ross, Hugh (1994), *Creation and Time* (Colorado Springs, CO: Navpress).

Roush, Wade (2000), "Pages–Book Reviews" ["Genome, Schmenone," A Review of Richard Lewontin's book, *It Ain't Necessarily So*], *Technology Review,* 103[3]:113, May/June.

Roth, Ariel A. (1998), *Origins: Linking Science and Scripture* (Hagerstown, MD: Review and Herald Publishing Association).

Rowan-Robinson, M. (1991), "Dark Doubts for Cosmology," *New Scientist,* 129:24-28, March 9.

Rupke, N.A. (1973), "Prolegomena to a Study of Cataclysmal Sedimentation," *Why Not Creation*, ed. Walter E. Lammerts (Grand Rapids, MI: Baker).

Sagan, Carl, ed. (1973), *Communications with Extra-terrestrial Intelligence* (Boston, MA: MIT Press).

Sagan, Carl (1974), "Life on Earth," *Encyclopaedia Britannica* (New York: Encyclopaedia Britannica, Inc.), 10:894ff.

Sagan, Carl (1979), "Will It All End in a Fireball?," *Science Digest*, 86[3]:8-15, September.

Sagan, Carl (1980), *Cosmos* (New York: Random House).

Sagan, Carl (1994), *Pale Blue Dot* (New York: Random House).

Sagan, Carl (1997), "Life," *Encyclopaedia Britannica* (New York: Encyclopaedia Britannica, Inc.), 22:964-981.

Salisbury, Frank (1969), "Natural Selection and the Complexity of the Gene," *Nature*, October 25.

Salisbury, Frank (1971), "Doubts about the Modern Synthetic Theory of Evolution," *American Biology Teacher*, 33:335-338, September.

San Diego Union (1982), "44% Believe God Created Mankind 10,000 Years Ago," August 30.

Sarfati, Jonathan D. (1998), "If God Created the Universe, Then Who Created God?," *Creation Ex Nihilo Technical Journal*, 12[1]:21.

Saunders, Will, et al. (1991), "The Density Field of the Local Universe," *Nature*, 349:32-38, January 3.

Schmidt, B., R. Kirschner, R. Eastman, M. Philips, N. Suntzell, M. Hamuy, J. Maza, and R. Aviles (1994), "The Distances to Five Type II Supernovae Using the Expanding Photosphere Method and the Value of *H*," *Astrophysics Journal*, 432:42-48.

Scott, Andrew (1985), "Update on Genesis," *New Scientist*, 106:30-33, May 2.

Shawver, Lisa J. (1974), "Trilobite Eyes: An Impressive Feat of Early Evolution," *Science News*, 105:72, February 2.

Sheler, Jeffery L. (1999), *Is the Bible True?* (San Francisco, CA: HarperCollins).

Silk, Joseph (1980), *The Big Bang* (San Francisco, CA: W.H. Freeman).

Silk, Joseph (1991), "Shedding Light on Baryonic Dark Matter," *Science*, 251:537-541, February 1.

Silk, Joseph (1992), "Cosmology Back to the Beginning," *Nature*, 356:741-742, April 30.

Simpson, A.J.G., F.C. Reinach, P. Arruda, et al. (2000), "The Genome Sequence of the Plant Pathogen *Xylella fastidiosa*," *Nature*, 406:151-156, July 13.

Simpson, George G. (1953), *The Major Features of Evolution* (New York: Columbia University Press).

Simpson, George G., C.S. Pittendrigh, and L.H. Tiffany (1957), *Life: An Introduction to Biology* (New York: Harcourt, Brace).

Simpson, George G., and William S. Beck (1965), *Life: An Introduction to Biology* (New York: Harcourt, Brace & World).

Slipher, Vesto M. (1913), "The Radial Velocity of the Andromeda Nebula," *Lowell Observatory Bulletin* No. 58, 2[8]:56-57.

Smolin, Lee (1995), "A Theory of the Whole Universe," in John Brockman (1995), *The Third Culture* (New York: Simon & Schuster), pp. 285-297 (with responses from other cosmologists on pp. 297-302).

Smoot, George and Keay Davidson (1993), *Wrinkles in Time* (New York: Avon).

Snelling, Andrew A. (1995), "The Whale Fossil in Diatomite, Lompoc, California," *Creation Ex Nihilo Technical Journal*, 9[2]:244-258.

Spetner, Lee M. (1997), *Not by Chance* (Brooklyn, NY: Judaica Press).

Sproul, R.C. (1992), *Essential Truths of the Christian Faith* (Wheaton, IL: Tyndale House).

Sproul, R.C. (1994), *Not A Chance* (Grand Rapids, MI: Baker).

Sproul, R.C., John Gerstner, and Arthur Lindsley (1984), *Classical Apologetics* (Grand Rapids, MI: Zondervan).

Stace, W.T. (1934), *A Critical History of Greek Philosophy* (London).

Steidl, Paul M. (1979), *The Earth, the Stars and the Bible* (Phillipsburg, NJ: Presbyterian & Reformed).

Stenger, Victor J. (1987), "Was the Universe Created?," *Free Inquiry*, 7[3]:26-30, Summer.

Sullivan, J.W.N. (1933), *The Limitations of Science* (New York: Viking).

Suzuki, David and Peter Knudtson (1989), *Genethics* (Cambridge, MA: Harvard University Press).

Taylor, Richard (1967), "Causation," in *The Encyclopedia of Philosophy*, ed. Paul Edwards (New York: Macmillan), 2:56-66.

Thaxton, Charles B., Walter L. Bradley, and Roger L. Olsen (1984), *The Mystery of Life's Origin* (New York: Philosophical Library).

Thompson, Bert (1977), *Theistic Evolution* (Shreveport, LA: Lambert).

Thompson, Bert (1989), "'Hopeful Monsters' and Evolution: Punctuated Equilibrium Examined," *Reason & Revelation*, 9:5-8, February.

Thompson, Bert (1995), *Creation Compromises* (Montgomery, AL: Apologetics Press).

Thompson, Bert (2000), *Creation Compromises* (Montgomery, AL: Apologetics Press), second edition.

Tipler, Frank (1994), *The Physics of Immortality* (New York: Doubleday).

Travis, J. (1994), "Hubble War Moves to High Ground," *Science*, 266:539–541, October 28.

Trefil, James (1984), "The Accidental Universe," *Science Digest*, 92[6]:53-55,100-101, June.

Tresmontant, Claude (1967), "It is Easier to Prove the Existence of God than It Used to Be," *Réalités*, April.

Tryon, Edward P. (1973), "Is the Universe a Vacuum Fluctuation?," *Nature*, 246:396-397, December 14. [NOTE: Tryon's article was reprinted in *Modern Cosmology and Philosophy* (1998), ed. John Leslie (New York: Prometheus), pp. 222-225.]

Tryon, Edward P. (1984), "What Made the World?," *New Scientist*, 101:14-16, March 8.

Twain, Mark (1883), *Life on the Mississippi* (Boston, MA: Gambit).

Van Flandern, Tom (2002), "The Top 30 Problems with the Big Bang," *Apeiron*, 9[2]:72-90, April.

Varghese, Roy A. (1984), *The Intellectuals Speak Out About God* (Chicago, IL: Regnery Gateway).

von Mises, Richard (1968), *Positivism* (New York: Dover).

Wade, Nicholas (2003), "Scientists Say Human Genome is Complete," *New York Times*, [On-line], URL: http://www.ny times. com/2003/04/15/science/15GENO.html, April 15.

Wald, George (1962), "Theories of the Origin of Life," *Frontiers of Modern Biology* (Boston, MA: Houghton-Mifflin).

Wald, George (1979), "The Origin of Life," *Writing About Science*, ed. Mary Elizabeth Bowen and Joseph A. Mazzeo (New York: Oxford University Press). [This is a reprint of Dr. Wald's original, award-winning article as published in *Scientific American*, August 1954.]

Watson, James D. and Francis H.C. Crick (1953), "Molecular Structure of Nucleic Acids: A Structure for Deoxyribose Nucleic Acid," *Nature*, 17:737-738.

Watson, Lyall (1982), "The Water People," *Science Digest*, 90 [5]:44, May.

Weinberg, Steven (1977), *The First Three Minutes* (New York: Basic Books).

Weisz, Paul B. and Richard N. Keogh (1977), *Elements of Biology* (New York: McGraw-Hill).

Wells, Jonathan (1998), "Unseating Naturalism: Recent Insights from Developmental Biology," *Mere Creation*, ed. William A. Dembski (Downers Grove, IL: InterVarsity Press).

West, R.R. (1968), "Paleontology and Uniformitarianism," *Compass*, Vol. 45, May.

Wheeler, John (1977), "Genesis and Observership," *Foundational Problems in the Special Sciences* (Dordrecht, Holland: Reidel).

Whitfield, John (2003), "Sharp Images Blur Universal Picture," *Nature*, [On-line], "Science Update," URL: http://www.nature.com/nsu/030324/030324-13.html, March 31.

"Whose Ape Is It, Anyway?" (1984), *Science News*, 125[23]:361, June 9.

Wicken, J. (1979), "The Generation of Complexity in Evolution: A Thermodynamic and Information-Theoretical Discussion," *Journal of Theoretical Biology*, 77:349-365, April.

Wilder-Smith, A.E. (1976), *A Basis for a New Biology* (Einigen: Telos International).

Wilder-Smith, A.E. (1987), *The Scientific Alternative to Neo-Darwinian Evolutionary Theory* (Costa Mesa, CA: TWFT Publishers).

Winchester, A.M. (1951), *Genetics* (Boston, MA: Houghton-Mifflin).

Wolff, M. (1994), *Exploring the Physics of the Unknown Universe* (Manhattan Beach, CA: Technotron Press).

Wysong, R.L. (1976), *The Creation-Evolution Controversy* (East Lansing, MI: Inquiry Press).

Zuckerman, Solly (1970), *Beyond the Ivory Tower* (New York: Taplinger).

[AUTHOR'S NOTE: I would like to gratefully acknowledge the research and writing assistance of Brad Harrub and Branyon May in the preparation of the material in chapter three on the Big Bang and related concepts.]